Stochastic Approaches for Systems Biology

Mukhtar Ullah • Olaf Wolkenhauer

Stochastic Approaches for Systems Biology

 Springer

Mukhtar Ullah
Department of Systems Biology
and Bioinformatics
Institute of Computer Science
University of Rostock
18051 Rostock
Germany
mukhtar.ulllah@uni-rostock.de

Olaf Wolkenhauer
Department of Systems Biology
and Bioinformatics
Institute of Computer Science
University of Rostock
18051 Rostock
Germany
olaf.wolkenhauer@uni-rostock.de

ISBN 978-1-4899-9491-2   ISBN 978-1-4614-0478-1 (eBook)
DOI 10.1007/978-1-4614-0478-1
Springer New York Dordrecht Heidelberg London

Printed on acid-free paper

Springer is part of Springer Science+Business Media (www.springer.com)

This work is dedicated to minds that are open to rational scientific reasoning and are not preoccupied with prejudices.

Preface

The discrete and random occurrence of chemical reactions, far from thermo-dynamic equilibrium, among less-abundant chemical species in single cells, necessitates stochastic approaches for modeling. Currently available texts on stochastic approaches relevant to systems biology can be classified into two categories. Books in the first category require the reader to have sufficient background of probability theory and focus directly on applications. Books in the second category take a two-step approach: first, they provide the necessary background in probability theory and then the concepts so developed are applied to model systems. We here follow the "introduce when needed" approach which is more natural and avoids distractions to the reader. While we still provide a review of probability and random variables, subsequent notions of biochemical reaction systems and the relevant concepts of probability theory are introduced side by side. This will hopefully lead to an intuitive presentation of the stochastic framework for modeling subcellular biochemical systems. In particular, we make an effort to show how the notion of propensity, the chemical master equation, and the stochastic simulation algorithm arise as consequences of the Markov property. The reader is encouraged to pay attention to this because it is not easy to see this connection when reading the relevant literature in systems biology. The nonobvious relationship between various stochastic approaches and our own struggle to find texts explaining them provided a motivation to write this book.

Throughout the text we use several examples to illustrate ideas and motivate stochastic modeling. It is shown how such systems can be studied with computer simulations, and the reader is encouraged to experiment with the programming code provided. Additionally, the cell cycle model is included as a more complex case study. Exercises in each chapter provide an opportunity to deepen one's understanding.

Another aspect of this work is a focus on analytical approaches. Most works concentrate on stochastic simulations: the exact stochastic simulation algorithm and its various improvements and approximations. This work is an attempt to complement those works on stochastic simulation approaches. The most common formulation of stochastic models for biochemical networks is the chemical master equation (CME). While stochastic simulations are a practical way to realize the CME, analytical approximations offer more insight into the influence of randomness. Toward that end, the two-moment approximation (2MA) is a promising addition to the established analytical approaches including the chemical Langevin equation (CLE) and the related linear noise approximation (LNA). The 2MA approach directly tracks the mean and (co)variance, which are coupled in general. This coupling is not obvious in CME and CLE and ignored by LNA and conventional differential equation models.

For the advanced reader, we include a chapter at the end that deals with a general Markov process with both continuous and jump character.

Readership

The most frequently used conceptual framework to model and simulate biological systems is that of differential equations. Their widespread use is also reflected by the fact that any undergraduate course in the engineering and physical sciences, as well as many life science programs, will teach ordinary differential equations and the basics of dynamical systems theory. The book *Computational Cell Biology* by Fall et al. [42] provides an excellent treatment of modeling with differential equations, including a short introduction to stochastic modeling. As we shall argue in this book, there are natural systems for which stochastic modeling is more appropriate. We would thus hope that an advanced undergraduate student would find the material of this book accessible and helpful. As a complementary text we recommend *Branching Processes in Biology* by Kimmel and Axelrod [82] as a textbook for stochastic approaches in biology. The book focuses on branching processes, in which an entity (e.g., cell or molecule) exists for a time and then may be replaced by one, two, or more entities of a similar or different type. The book provides an excellent introduction to the theory but at the same time provides various examples that are relevant to experimentalists. Examples and application areas include the application of branching processes to polymerase chain reactions, DNA sequence analysis, cell cycle kinetics, drug resistance, and chemotherapy. With these examples and the combination of theory and practical examples it is a suitable complement for further reading to the present text.

We are well aware of the fact that stochastic modeling is less frequently, and less thoroughly covered by university courses. We admit that the material appears more abstract at first sight and takes some getting used to. We therefore encourage the reader of this book to experiment with the examples provided. To this end, we have included the code with which one can simulate these systems.

Computer Experiments

Stochastic modeling requires some of the more advanced material a graduate student encounters, and the notation alone can be daunting. To ease the pain, we have included a glossary and various examples, supported by the code that can be used to reproduce the examples. There are numerous software tools available to model and simulate dynamical systems. For most of the examples in this book we use Matlab from MathWorks [96], including the SimBiology and Symbolic toolboxes. An alternative software, which we have also used, is Cains http://cain.sourceforge.net, a free tool that excels in computationally efficient stochastic simulations. Anyone looking for more examples of subcellular biochemical networks for modeling will find model databases such as BioModels http://www.ebi.ac.uk/biomodels-main/ and JWS Online http://jjj.biochem.sun.ac.za useful.

Supporting material for this book, including software and a list of corrections, will be provided on our website at www.sbi.uni-rostock.de

Acknowledgements

The book documents a friendship between the authors that began in 2003, when we both moved from England to Germany, forming the nucleus of the new systems biology research group at the University of Rostock. The first couple of years were stressful, and our regular scientific discussions provided a sanctuary and escape from things happening around us. Our discussions had no other purpose than to improve our understanding of a problem under consideration. The mixture of competing ideas and collaborative effort allowed us to derive pleasure from proving ourselves *wrong*. Although we were interested in a wide range of questions related to systems biology, the theme of "uncertainty" appeared in various disguises. This eventually culminated in Mukhtar's PhD thesis, which provided the starting point for this book. Stochastic modeling is a wide area with a long history in the natural sciences and physics in particular. Knowing our own limitations, we decided on a concise form, tracing our discussions over the last four years, rather than attempting the seemingly impossible task of writing a comprehensive account of existing stochastic approaches. We would very much appreciate feedback, corrections, and suggestions on the text.

During the writing of this book we received support from various sources, including the Deutsche Forschungsgemeinschaft (DFG), the German Federal Ministry for Education and Research (BMBF), the Helmholtz Gesellschaft, the European Commission, and the Stellenbosch Institute for Advanced Study (STIAS). We are grateful for the financial support and feel slightly embarrassed that our work has been such an enjoyable experience. We thank the following people for their willingness to discuss the work described in this book: Thomas Millat (University of Rostock), Hanspeter Herzel (Humboldt University, Berlin), Akos Sveiczer (Budapest University of Technology and Economics), Kevin Burrage (University of Queensland, Brisbane), Allan Muir (Llanfynwydd, Wales) and Johan Elf (Uppsala University, Sweden).

In October 2010, Allan Muir, a good friend and twice a visitor to the
Department of Systems Biology & Bioinformatics at the University of Rostock,
died of cancer. For most of his academic working life Allan Muir was associated
with the Department of Mathematics at City University, London, where he
worked mostly on game theory. He never retired from doing mathematics;
his interest in math, science, and philosophy were inseparable from himself.
His astonishing general knowledge, his unique style of approaching a problem
through questions and thus reducing something complex to its essence, could
connect seemingly unrelated things and communicate the result to the lucky
ones who were allowed to tune in. He will be deeply missed.

Olaf Wolkenhauer (olaf.wolkenhauer@uni-rostock.de)
Mukhtar Ullah (mukhtar.ullah@uni-rostock.de)

Rostock, May 2011

Contents

Acronyms

ODE	ordinary differential equation
CME	chemical master equation
CLE	chemical Langevin's equation
SDE	stochastic differential equation
SSA	stochastic simulation algorithm
LNA	linear noise approximation
FPE	Fokker–Planck equation
CKE	Chapman–Kolmogorov equation
dCKE	differential Chapman–Kolmogorov equation
2MA	two-moment approximation
RC	reaction count
CDF	cumulative distribution function
CCDF	complementary CDF
PDF	probability density function
PMF	probability mass function
MPF	metaphase promoting factor
CT	cycle time
BM	birth mass
DM	division mass
CV	coefficient of variation
NSR	noise-to-signal ratio
xNSR	cross noise-to-signal ratio
SD	standard deviation

Notation

Conventions

We shall distinguish three ways of highlighting terms. A term (i) we wish to emphasize, (ii) that appears for the first time, or (iii) is defined implicitly as part of a sentence will be typeset in italics. When we inform the reader of how a term is referred to (commonly called) in the literature, we enclose it in double quotes. When we ourselves introduce a new term, that is, we name something in our own way, we enclose it in single quotes.

General Notes

- Symbols of the form $Q(t)$ represent time-dependent quantities.

- The ith element of a vector X is denoted by X_i.

- For a matrix S, the transpose is written S^T

- The entry that lies in the ith row and the jth column of a matrix S is denoted by S_{ij}. When a complete row/column is intended, we write a dot for the index of the column/row. Thus we write $S_{i.}$ for the ith row, and $S_{.j}$ for the jth column.

- Any symbol denoting a random/stochastic variable/process will be set as an uppercase letter. Once defined, the corresponding lowercase symbol represents a sample/realization of the variable/process. The time-dependent version of the same symbol in lowercase represents a deterministic approximation of the corresponding stochastic process. Thus if $N(t)$ is a stochastic process, n is a typical sample of $N(t)$ and $n(t)$ is a deterministic approximation of $N(t)$.

- Symbols denoting operators such as probability are typeset in roman, e.g., $\Pr[\cdot]$.

- The notation $f(n)$ is shorthand for $f(n_1, \ldots, n_s)$ whenever an s-vector n appears as an argument.

- We will switch between the three alternative notations $\dot{\phi}(t)$, $\frac{\mathrm{d}}{\mathrm{d}t}\phi(t)$ and $\frac{\mathrm{d}\phi}{\mathrm{d}t}$, for the time derivative of any scalar quantity $\phi(t)$. We will prefer the last of these when dependence on time is implicitly clear.

List of Symbols

X_i	ith chemical species/component
$\varnothing$	null species
s	number of chemical species
R_j	jth reaction channel
r	number of reaction channels
$N_i(t)$	copy number
$N^c(t)$	continuous approximation of $N(t)$
N_A	Avogadro's constant
V	volume
Ω	system size, typically $N_A V$
Ω	sample space
$X_i(t)$	species concentration (absolute or relative)
$C_i(t)$	characteristic concentration indicating the scale of $X_i(t)$
$Y_i(t)$	general stochastic process
$\phi(t)$	concentration for an infinitely large system size
$\Xi(t)$	fluctuation of $N^c(t)$ around $\phi(t)$
ξ	a sample (realization) of $\Xi(t)$

S	stoichiometry matrix
$\underline{S}_{ij}$	number of X_i molecules as reactant in channel R_j
$\bar{S}_{ij}$	number of X_i molecules as product in channel R_j
$Z_j(t)$	reaction count: number of R_j occurrences during $[0, t]$
$\Pr[\cdot]$	probability measure
$P(n, t)$	state probability
$P^c(n, t)$	continuous approximation of $P(n, t)$
$\Pi(\xi, t)$	$P^c\left(\Omega\phi(t) + \Omega^{1/2}\xi, t\right)$
$\delta(n)$	Dirac's delta function centered at the origin
$P(n\|m, t)$	transition probability
$v_j(x)$	reaction rate
k_j	rate coefficient, or rate constant
$a_j(n)$	reaction propensity
c_j	stochastic reaction rate constant
$h_j(n)$	different possible combinations of R_j reactant molecules
$\hat{v}_j(n)$	conversion rate
$a_0(n)$	exit rate, $\sum_j a_j(n)$
$\mathbb{E}_j$	a step operator defined by $\mathbb{E}_j f(n) = f(n - S_{\cdot j})$
$\langle Y(t)\rangle$	mean of the process $Y(t)$
δY	deviation from the mean, $Y - \langle Y\rangle$
$\langle\delta Y\delta Y^T\rangle$	covariance matrix of the process $Y(t)$

$\langle \delta Y_i \delta Y_k \rangle$ (i,k)th element of the above covariance matrix

$\langle \delta Y_i^2 \rangle$ variance of the process Y_i

$\mu_i(t)$ mean concentration $\langle X_i(t) \rangle$

$\sigma_{ik}(t)$ pairwise concentration covariance $\langle \delta X_i \delta X_k \rangle$

$\zeta_{ii}(t)$ NSR

$\zeta_{ik}(t)$ xNSR

$A(n)$ drift rate of (the process) $N(t)$ in state n

$B(n)$ diffusion rate of (the process) $N(t)$ in state n

$f(x)$ drift rate of (the process) $X(t)$ in state x

$g(x)$ diffusion rate of (the process) $X(t)$ in state x

$A : B$ Frobenius inner product: sum of elements of $A \odot B$

$\mathcal{P}_j$ Poisson random variable

$\mathcal{N}_j$ standard normal random variable

$\mathcal{W}_j$ standard Brownian motion, or Wiener process

$T_j(n)$ time, in state n, until the occurrence of a reaction R_j

W_k The kth waiting time in a counting process

T_k interarrival time between the kth and the next arrival

$J(n)$ index of the next reaction known to have occurred

$\mathcal{O}(x)$ first neglected order with respect to x in an expansion

$o(x)$ terms vanishing faster than x, as the latter approaches zero

$X \sim Y$ Random variables X and Y are identically distribed

$\mathrm{diag}(a)$ diagonal matrix with elements of vector a on the main diagonal

$\#A$ number of elements in a set A

List of Figures

List of Tables

List of Matlab Functions

List of Revisited Examples

Chapter 1

Introduction

1.1 Levels of Organization in Living Systems

All known living systems are made up of one basic structural and functional unit: the cell. Populations of interacting cells can form higher levels of structural and functional organization. Take, for example, the human body, in which cells are organized to form tissues and organs. Cells in tissues and organs are specialized for a particular function they realize within that context. One can thus distinguish between the molecular level (molecules interacting within the cell), the cellular level (interacting cells), and the level of tissues or organs. Various organs together make up an organ system, and various organ systems together make up an organism. Now, with a complex system such as the human body, the different levels of structural and functional organization are tightly coupled. This is illustrated in Figure 1.1, which shows the human body as a multilevel system. The digestive system includes the small and large intestines in which food is processed. The large intestine is further divided into the cecum and colon. The colon is a common site of carcinogenesis due to the mechanical and chemotoxic stress it is subjected to. It is for this reason the object of intensive research. The functional role of the colon is thus the absorption of nutrients from food that passes through the lumen. The lumen is the inner tract of the intestine tract, further organized into villi (vaginations, folds in the small intestine) and crypts (cavities) that effectively increase the overall surface area. The innermost tissue lining consists of cells to absorb nutrients and goblet cells that secrete mucus, which lubricates the passage of the food through the intestine. There are about 10^7 crypts, each consisting of several thousand cells. At the bottom of the crypt, a small number of stem cells divide slowly in an environment referred to as the niche. The functional role of the stem cell niche is to renew the tissue on a weekly basis and to repair damaged tissue. The daughter cells of dividing stem cells proliferate rapidly before differentiating and maturing into functional tissue cells. The cells of the crypt walls migrate toward the top, where they undergo apoptosis (cell death) and/or are shed into the gut lumen. Homeostasis of the overall system involves various levels of structural and functional organization. Each level possesses a characteristic function and interlinked dynamic behavior, regulated through

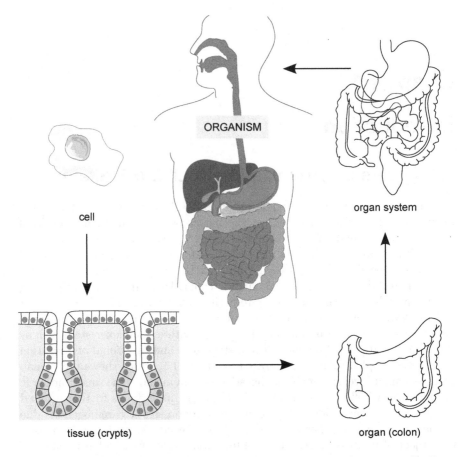

Figure 1.1 Cells—the building blocks of living systems. Each type of cell is specialized for a particular function. Large numbers of specialized cells together make up a tissue. Various tissues together make up an organ. Various organs together make up an organ system, and various organ systems together make up an organism.

cellular processes. The goal of systems biology is to understand the behavior of such biological systems through an understanding of cells.

The most obvious aspect of the structural organization of the (eukaryotic) cell (Figure 1.2) is given by the outer membrane and the inner membrane, which defines the *nucleus*. While the *prokaryotic cell* (microorganisms, bacteria, etc.) is characterized by only one compartment, the *eukaryotic cell* has the inner membrane that defines the nucleus. The nucleus of the cell contains the genetic material, or *genome*, in the form of a double-stranded DNA molecule

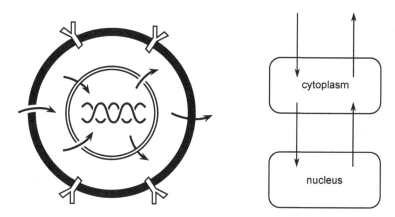

Figure 1.2 Structural organization of the (eukaryotic) cell. The defining boundary structures are the outer membrane, and the inner membrane that defines the nucleus. Material, signals, and information can pass through the membranes directly, through gates, or through receptors. What the drawing does not show are two important structural elements: the cytoskeleton (providing structural support and transport mechanisms) and other organelles in the cytoplasm that fulfill specialized roles in the processing of proteins.

with its characteristic double helix structure. The genetic material is packed into *chromosomes*. The generic term *gene* is used to describe the role of information- and protein-coding regions in the genome. The medium between the nucleus and the outer membrane is the intracellular fluid *cytosol*. The area between the outer and inner membranes, including all of the components therein, is called *cytoplasm*. The *cytoskeleton* is a meshwork providing structural support for the cell. As part of the cytoskeleton, *microfilaments*, made of *actin*, provide mechanical support (and participate in some cell–cell or cell–matrix interactions), while *microtubules*, made of *tubulin*, act as a transport system for molecules. In addition to the two main compartments (nucleus and cytoplasm), eukaryotic cells have *organelles*, which are smaller compartments with a membrane and which contain a set of specific enzymes. Material can pass through the membranes directly or through pores. More specifically, there are four kinds of proteins that are embedded in the outer cell membrane and organelle membrane to allow material import and export: pores, ion channels, transporters, and pumps. In contrast to the structural organization of the cells, its functional organization involves the following key processes, or *cell functions*:

- Growth

- Division (cell proliferation)

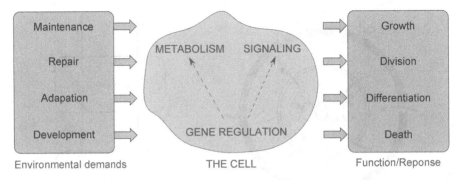

Figure 1.3 The functional organization of the cell.

- Specialization (cell differentiation)

- Death (apoptosis)

In order to investigate the (mal)functioning of cells, cell populations, tissue, organs, and organisms, one has to gain an understanding of the behavior of cells—processes therein and interactions between cells and their environment.[1] This quest can be summarized by the following two questions:

Intracellular dynamics: How do the components within a cell interact to form the cell's structure and realize its function? This is an interior aspect of the cell.

Intercellular dynamics: How do cells interact to develop and maintain higher levels of structural and functional organization? This is an exterior aspect of the cell.

To this end, we can group processes within a cell as follows (Figure 1.3):

- *Gene regulation*: The reading and processing (transcription and translation) of information from the genome in response to the environment.

- *Metabolism*: Processes that construct and maintain the cell, that realize cell growth and genome duplication before cell division.

- *Signal transduction*: Processes of inter- and intracellular communication and the coordination of cell function.

Each class of processes usually involves a large number of interacting molecular species organized in networks (pathways). The threefold classification of

---

[1]The environment of a cell consists of other cells and the extracellular matrix.

cellular processes into metabolism, signaling, and gene regulation associates with each a range of specialized technologies for generating experimental data. The nature of the data can differ considerably, making their integration a challenge. At the methodological level, where one is trying to model and simulate these processes, a range of approaches is used, depending the type of network under study. The biochemical reactions of metabolism are organized into metabolic pathways, whereas reaction networks underlying signaling are organized into signal transduction pathways. Genes are sometimes regarded as nodes in a gene regulatory network, with inputs being proteins such as transcription factors, and outputs being the level of gene expression.[2]

The study of metabolism, cell signaling, and gene expression requires a range of technologies, often leading to an operational division of researchers into "omics" disciplines, including metabolomics, proteomics, and transcriptomics. While there is an obvious relation between metabolism, signaling and gene expression, the complexity of the cell, specifically the technological difficulties of measuring these processes, has forced researchers to specialize with obvious consequences for the overall endeavor—we can't see the forest wood for the trees.

While to this day the identification and molecular characterization of cellular components has been the main focus in molecular and cell biology, the emergence of systems biology is closely linked to the fact that the functioning of cells is an inherently dynamical phenomenon. In our view, systems biology is thus the timely merger of (dynamical) systems theory with molecular and cell biology.

The modeling of dynamical systems, and biochemical reaction networks in particular, can be roughly divided into three classes:

- Models based on differential equations

- Stochastic models

- Other approaches, such as Petri nets, pi-calculus, and combinations of methodologies.

Differential equations are the most frequently used approach to representing dynamical systems. In systems biology, they derive their popularity from an apparently direct translation of biochemical reactions into rate equations. While the mathematical structures of such rate equations are often similar or identical, the semantics can differ widely (See Figure 2.4 in Chapter 2). This highlights an important fact: choosing a modeling framework depends on the nature of the system under consideration but also on assumptions and personal preferences.

---

[2]A glossary in the appendix provides brief definitions for key biological terms.

As a consequence of the complexity of biological systems, full understanding of cells and their function(ing) cannot be assured. Hypotheses must thus be formulated and tested by experiments. This requires a conceptual framework appropriate for making precise and empirically testable predictions. Such a framework is provided by (dynamical) systems theory. A mathematical model is a thus representation, a simplified version of the part of the biological system studied, one in which exact calculations and deductions are possible. An obvious priority in modeling is assurance that the model's behavior (established through numerical simulation or formal analysis) corresponds closely to the empirical behavior of the biological system, that the mathematical model in some way resembles the behavior of the biological system. In addition to replication/reproduction of certain observed qualities or behavior, simplicity and mathematical tractability can be important criteria in developing a model.

A biological system is our interpretation of observable facts in the light of a formal model that we ourselves invent/construct. Understanding a complex system thus requires abstraction, reducing one type of reality to another. Mathematical modeling facilitates understanding through abstraction. If we are to describe the mechanisms/principles/laws by which the components of a system interact (and thereby realize the (sub)system functions), then the purpose of the model is to distill something complex to a simpler, essential aspect. Modeling does therefore imply for most cases a reduction of complexity; a model is then understood as an excerpt or selection from the biological system under consideration.

To model inter- and intracellular processes, one requires quantitative spatiotemporal data for a relatively large number of components. At present these are not available, forcing us to handle uncertainty and "reduce" complexity. For practical purposes to do with technological limitations, but also with the time and money required to conduct the experiments, a subset of components is chosen. This leads to the pragmatic notion of pathways or networks as a selected subsystem of biochemical reactions (relevant to some cell function).

Not only are we forced to select a subset of proteins, respectively a subsystem, even if we could quantify larger numbers of components, the analytical tools for the analysis of such large, nonlinear models are missing. Proteins are modified (e.g., activated), each of these states adding to the number of variables in a mathematical model. A system with 10 components can subsequently lead to 20 or more system variables. The theory of nonlinear dynamical systems, the methodologies and tools available to identify models (their structure and parameter values) from experimental data, to investigate their behavior analytically or through numerical simulations, remains to this day limited. We are once more forced to simplify out of practical considerations. The reduction of complexity through abstraction and modeling does, however, serve more than that. In studying complex systems, we seek simplifications to reduce complex processes to an essential aspect of their functional organization, to extract a principle that serves as an explanation. We are seeking general principles underlying the observations we make in experiments. Mathematical modeling is then the art of making appropriate assumptions, balancing necessary reductions due to methodological and experimental limitations with abstractions serving explanatory purposes.

The functions of a cell are thus realized by spatiotemporal processes. The first omission we admit here is that we will largely ignore spatial aspects. Within cells, the translocation of molecules can either be assumed to be so rapid that it does not matter or, if barriers are crossed (say the inner nuclear membrane of eukaryotic cells), we might approximate this translocation process by a reversible reaction. Spatial aspects are important, and we shall not pretend otherwise.

In the present text we shall look at cells as the building blocks of living systems. Observing cells in experiments, irregularities and the absence of an obvious pattern/trend in data induce uncertainty in the analysis of the system. The first question is then whether this randomness is an inherent, possibly purposeful aspect of the system or whether it is a consequence of limitations in observing the system (the choice of subsystem looked at, ignored components or limitations to measurement technologies)?

Note that our discussion will also be limited to the level of cells, where we investigate the function(ing) of cells in terms of changes in the abundance of molecules within cells and consequences this may have for populations of interrelated cells [126]. The discussion of randomness in physics, specifically statistical mechanics, may thus be avoided in our present context. While thermal and perhaps quantum fluctuations influence events at the cellular level and above, instead of modeling them in detail we may, without losing essential cellular and higher-order modeling power, represent their consequences by irreducible stochasticities. The cell is here considered an open, nonequilibrium system, with a constant flux of material and information into and out of the cell. At the level of single molecules, the irregular motion of atoms and molecular bonds within the system may well be relevant but will here be referred to as effects of a 'microscopic level'. This includes thermal fluctuations and *Brownian motion*. Looking at changes in the concentration of molecules, following a clear trend that can be described in terms of differential equations, such models may be referred to as 'macroscopic'. Our focus will here be the level of changes in the population/concentration of molecules, without further consideration of the mechanistic details underlying the reactions.

1.2 Systems Biology

Systems biology takes an interdisciplinary approach to the systematic study of complex interactions in biological systems. This approach seeks to decipher the emergent behaviors of complex systems rather than focusing only on their constituent properties. Aiming at understanding the dynamic interactions among components of a cell, and among cells as well as their interaction with the environment, systems biology is an approach by which biomedical questions are addressed through integrating experiments in iterative cycles with mathematical modeling, simulation, and theory. Modeling is not the final goal, but is a tool to increase understanding of the system, to develop more directed experiments, and finally to enable predictions. Mathematical models have the advantage of being quantitative and interactive rather than solely descriptive. The process by which models are formulated, which may include the representation of genetic, epigenetic, cellular, and tissue effects across the various physical and temporal scales during tumorigenesis, helps to articulate hypotheses and thereby supports the design of appropriate experiments to test them [170].

The most popular definitions of systems biology refer to dynamics, mechanisms, principles, and behaviors. The complexity of biological systems and/or functions arises from the interaction of myriad nonlinear spatiotemporal phenomena and components. The fact that most cellular processes, such as cell-cycle control, cell differentiation, and apoptosis, are inherently dynamical highlights the need for integrating mathematical modeling into life science and clinical research. A systems biology approach can help identify and analyze the principles, laws, and mechanisms underlying the behavior of biological systems.

In systems biology, arguments arise over the predictions and validity of theories, the methods of collecting data, and the interpretation of experimental data sets. Figure 1.4 describes the role of mathematical modeling within the field of systems biology, mediating the interpretation of experimental data and helping the formulation of hypotheses. A warrant is the justification that explains the relation of the data to the hypothesis (claim). Often, warrants rest on contextual assumptions that are only tacitly acknowledged. Qualifiers express the limits of the validity of the claim. Arguments arise when attempts are made to rebut or refute the claim either by attacking the validity of the data or the validity of the warrant. The diagram shows how mathematical modeling fits in Toulmin's philosophy of argumentation.

The main reason that necessitates modeling in the life sciences is the complexity of natural systems. The number of components does not really play a particular role in this. To have many molecules or cells interacting is not a problem as such (particularly not if they are in sync or if it is an

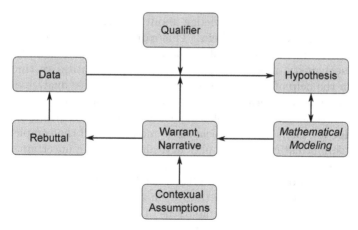

Figure 1.4 The role of mathematical modeling in systems biology. Figure derived from [111] and based on Toulmin's theory of argumentation [153].

average process that matters). The number of different kinds of components does, however, provide a challenge for the theory of nonlinear systems, which to this day is practical only for a handful of system variables. The fact that a molecule and its modified form (say by phosphorylation) require already two system variables in a model of differential equations shows that systems with ten or twenty molecular species can become rather difficult to handle. One should also add the difficulties in accounting for spatial phenomena. While models based on ordinary differential equations dominate, it is at present not practical to formulate partial differential equation models and identify their parameters from experimental data. Nonlinearity in the interactions and the resulting behavior is another major element of complexity, a hurdle and a source for surprise. Most important, however, is the fact that a natural system constantly changes, adapts, evolves, making it difficult to use approaches that assume stationarity and time-invariance. A living system, such as a cell, a tissue, an organ, or an organism, is acting and reacting; it responds to and modifies its environment. Cells, like organisms, undergo a continuous process of mutual interaction and change. A living system is constrained by its environment but also changes its environment. This self-reference and the subsequently emerging phenomena are the real cause of trouble for the modeler.

While mathematical modeling (especially of dynamical systems) is a central element of systems biology, the field of bioinformatics has more to do with the analysis of data and information, whether directly from experiments or from databases and the literature. Both areas are complementary and rely on each other. For example, to simulate a signaling pathway, the construction of the model benefits from knowledge about the structural properties, e.g., phosphorylation sites, of a protein. This can be found in databases using bioinformatics methods and tools. For many systems we cannot develop detailed mechanistic models because of a lack of quantitative time-course data. With the help of bioinformatics one can nevertheless analyze databases to help formulate and verify hypotheses about networks at a higher level of abstraction. Although the present text focuses on systems biology, one should acknowledge the challenges of mechanistic modeling and the complementary role that bioinformatics methods play in dealing with the uncertainty arising from the complexity of the systems under consideration.

There are two dominant paradigms used in mathematical modeling of biochemical reaction networks (pathways) in systems biology: the deterministic approach, using numerical simulations of nonlinear ordinary differential equations (including mass-action-type, power-law or Michaelis–Menten models), and the stochastic approach based on a master equation and stochastic simulations. Stochastic modeling has a long tradition in physics and has brought forth such expository masterpieces as the books by van Kampen [75] and Gardiner [47]. Van Kampen shows how a stochastic model can be formulated comprising both the deterministic laws and the fluctuations about them. Such models are sometimes referred to as "mesoscopic" models. Considering a system of interacting mass points, fluctuations in nonequilibrium systems do not arise from a probability distribution of the initial microstate, but are continuously generated by the equations of motion of the molecules. While mesoscopic stochastic models are attractive theoretical concepts, in a practical context where such a (nonlinear) model and its parameter values would have to be extracted from experimental data, we face various problems (which are in part a reason for the wide use of ordinary differential equations).

We can illustrate the notions of microscopic, mesoscopic, and macroscopic in the context of cell biology by considering gene expression, the process by which information of the genome is first transcribed into RNA before being translated into proteins. These two stages involve two levels, the transcription of a gene being microscopic compared to fluctuations in the concentration of the protein for which the gene encodes the information. While for the initiation of transcription, say through the binding of transcription factors, a discrete stochastic model may be appropriate, changes in the concentrations of the proteins involved in the function of a single cell (e.g., cell cycle) may on the other hand be described macroscopically by ordinary differential equations. That, however, is only valid if effects of discrete random events at the level of transcription do not propagate to the level of translation. When that happens, the continuous model describing changes in protein concentrations needs to be stochastic as well and will take the form of a so-called "Langevin equation."

In many situations random fluctuations are sufficiently small to be ignored, allowing macroscopic equations to predict the behavior of a system with great accuracy. Cells, however, are *open systems*, where the environment may force them into a stationary nonequilibrium state in which the system's dynamics bifurcate, the direction taken depending on the specific fluctuations that occur. Note that therefore the randomness of the fluctuations (which can be described only in terms of probabilities) influences the behavior of the system of macroscopic equations most critically at specific bifurcation points, while other areas of the state space may be perfectly well approximated by macroscopic equations. Intrinsic noise from thermal fluctuations or transcriptional control could determine how the system at the macroscopic level goes through a bifurcation. Looking at a population of genetically identical cells in a homogeneous environment, this leads to variability of cell states that may well be exploited by the biological system [74, 127, 141]. The obvious context in which randomness has a function is generating diversity in evolution.

Looking at a single gene in a single cell, the initiation of transcription at its promoter site is driven by the association and dissociation of a very small number of molecules. This very low copy number of molecules has two consequences: the time of reaction events can be described only in terms of probabilities, and changes in the number of molecules are discrete, with no obvious trend that could be approximated with a differential equation (see [117] for a review). The expression of a gene does, however, serve a cell function such as growth, differentiation, and apoptosis. For example, in response to external stimuli, the cell may produce large quantities of a protein. This response, measured as an apparently smooth/monotonic change in concentration, is appropriately described by differential equations. Small fluctuations around an obvious trend/mean are thus ignored. At this level we are aiming at a description of a pathway acting as a switch, filter, oscillator, or amplifier, studying the network behavior in terms of its robustness, responsiveness, and sensitivity of the model to changes in parameters, transitions between steady states, and bifurcations. A usual assumption in such rate equation models is that parameters (rate coefficients) are constants. Since these parameters are implicitly linked to environmental variables, such as temperature, pH level, or water balance, fluctuations in these are considered negligible. The art of modeling is then to decide in the given context which modeling approach or combination of approaches is most appropriate.

The following section will serve as additional motivation for stochastic modeling.

1.3 Why Stochastic Modeling? Philosophical

Two opposing views of causal entailment in nature exist: *determinism* and *randomness*. Adherents of determinism assert that events in nature are governed by causal laws that unambiguously link changes of relevant quantities (states). That is, the next state of the system is unambiguously determined by the present and past states of the system. Adherents of randomness, on the other hand, assert that nature is random at a fundamental level and the course of future events cannot be predicted from the knowlegdge of previous events in a deterministic sense. In systems biology, a modeler should not worry whether nature is deterministic or random at a fundamental level. The choice of a modeling framework is determined by the complexity of the system being modeled, the level of investigation, and, consequently, the question being asked [29]. Even under the assumption of determinism as a worldview, successful modeling of complex phenomena requires stochasticity, owing to the complexity of the natural system as a whole. Each observable phenomenon is causally related to a large number of phenomena, and its pattern of development depends on many factors, not all of which can be established and traced. For this reason, each observation of the same phenomenon shows, besides its general properties, certain special features that are typical only of that particular observation. These observation-specific features, referred to as fluctuations, are products of the influence of the factors excluded from the model. This loss of information is handled by probabilistic methods. Thus, the assumption of a stochastic model is a scientific decision, not a metaphysical perspective. Andrei Kolmogorov, the founder of modern probability theory, explains the matter in his statement that "the possibility of applying a scheme of either a purely deterministic or a stochastically determined process to the study of some real processes is in no way linked with the question whether this process is deterministic or random" [83]. The so-called "real process" is not accessible to scientific investigation. Whether a cell function is deterministic or stochastic cannot be answered in the current realm of science, and even if cell function were deterministic, that would not be reflected in a gene network, since the genes in the model would be affected by events (latent variables), including genes, outside the model, thereby forcing the modeler to choose a stochastic model. This recognition is critical in modeling (intra- and inter-) cellular phenomena.

1.4 Why Stochastic Modeling? Biological

At a coarse level, cell functions are largely determined by spatiotemporal changes in the abundance of molecular components. At a finer level, cellular events are triggered by discrete and random encounters of molecules [120]. The discreteness is typical of processes with only a few molecules. Gene transcription is an example of such discrete processes. That cellular events are discrete and random is supported by many recent experiments [7, 39, 97, 127] that have revealed cell–cell variations, even in isogenic cell populations, of

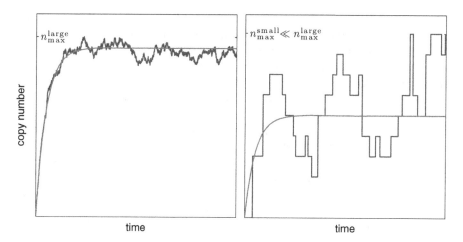

Figure 1.5 Discrete and random nature of chemical reactions. *Left*: large copy numbers and frequent reactions allow for a continuous approximation leading to the chemical Langevin equation (Chapter 5), which, for an infinitely large system, approaches deterministic rate equations. *Right*: small copy numbers and infrequent reactions require discrete stochastic approaches leading to the chemical master equation and stochastic simulations.

transcription (when a gene is copied in the form of an mRNA transcript) and translation (when the mRNA is used as a template to build proteins).

The above discussion may suggest a deterministic modeling approach at the coarse level (cell function) and a stochastic one at the finer level (gene regulation) [6, 13, 74, 93, 95, 115, 116, 122, 127, 128]. However, stochastic modeling is necessary when noise propagation from processes at the fine level changes cellular behavior at the coarse level.

Stochasticity is not limited to low copy numbers. The binding and dissociation events during transcription initiation are the result of random encounters between molecules [74]. If molecules are present in large numbers and the molecular events occur frequently, as in Figure 1.5 (left), the randomness will cancel out (both within a single cell and from cell to cell) and the average cellular behavior could be described by a deterministic model. However, many subcellular processes, including gene expression, are characterized by infrequent (rare) molecular events involving small copy numbers of molecules, as in Figure 1.5 (right) [74, 120]. Most proteins in metabolic pathways and signaling networks realizing cell functions are present in the range 10–1000 copies per cell [14, 91, 117]. For such moderate/large copy numbers, noise can be significant when the system dynamics are driven toward critical points in cellular systems that operate far from equilibrium [34, 150, 176]. The significance of noise in such systems has been demonstrated for microtubule formation [28], ultrasensitive modification and demodification reactions [14], plasmid copy number control [119], limit cycle attractor [125], noise-induced oscillations near a macroscopic Hopf bifurcation [162], and intracellular metabolite concentrations [37].

Noise has a role at all levels of cell function. Noise, when undesired, may be suppressed by the network (e.g., through negative feedback) for robust behavior [46, 101, 118, 127, 132, 151]. However, all noise may not be rejected, and some noise may even be amplified from process to process and ultimately influence the phenotypic behavior of the cell [13, 69, 86, 122, 142]. Noise may even be exploited by the network to generate desired variability (phenotypic and cell-type diversification) [16, 24, 64, 127, 173]. Noise from gene expression can induce new dynamics including signal amplification [133], enhanced sensitivity (stochastic focusing) [121, 122], bistability (switching between states) and oscillations [8, 9, 45, 92, 112], stabilization of a deterministically unstable state [154], and even discreteness-induced switching of catalytic reaction networks [152]. These are both quantitatively and qualitatively different from what is predicted or possible deterministically. Other important processes wherin noise plays a role include development and pattern formation [88].

In the remainder of the present section, we illustrate the need for stochastic modeling by selecting a few important aspects of biochemical reaction networks.

Identifiability: In the isomerization reaction $U \overset{k_w}{\underset{k_u}{\rightleftharpoons}} W$, proteins are converted back and forth between the unmodified form U and the modified form W, such that the total number n^{tot} of protein molecules remains constant. When treated deterministically, the number n of proteins in the unmodified form varies continuously with time according to the rate equation, to be derived in the next chapter,

$$\frac{\mathrm{d}n}{\mathrm{d}t} = k_u n^{\text{tot}} - (k_w + k_u)n,$$

where k_w and k_u are the respective rate constants of the modification and demodification reactions. When recast in nondimensional time τ, the rate equation takes the form

$$\frac{\mathrm{d}n}{\mathrm{d}\tau} = \frac{k_u n^{\text{tot}}}{(k_w + k_u)} - n.$$

Here we see that $n(\tau)$ depends on the fraction $p_u = k_u/(k_w + k_u)$ but not on the particular values of k_w and k_u. In other words, experimental data on protein copy numbers can provide information only about the fraction p_u, and not on the particular values of k_w and k_u separately. This issue of identifiability is reported in [165, 166]. The problem here is that changes in the protein copy numbers are discrete and random, rather than continuous and deterministic. We will learn in Chapter 6 that the variance of n satisfies an ordinary differential equation (ODE) that involves the difference $k_w - k_u$ between the two parameters, in addition to the fraction p_u. Thus experimental data on fluctuations, combined with the experimental data on the average protein copy numbers, give information about both k_w and k_u separately. Note that this argument of identifiability is made in the context of parameter estimation from time courses. Some experimental procedure may allow one to identify both parameters directly without requiring estimation from time course.

Depletion (extinction): The steady-state copy number, n^{ss} in the above example, is a fraction of the total copy number n^{tot} and hence can never become zero for nonzero rate constants. However, a discrete stochastic treatment of the problem leads to nonzero steady-state probabilities of $n = 0$ (corresponding to depletion of U), and of $n = n^{tot}$ (corresponding to depletion of W). Of specific interest in such cases is the average time until the first depletion (or extinction).

Fluctuation-affected mean: The validity of deterministic macroscopic approaches for description of the averages is limited because the average of a nonlinear function is generally not the same as the function of the average. This was first demonstrated for bimolecular reactions in [131]. In the isomerization example, the mean (copy number) of the stochastic model was the same as the solution of the corresponding deterministic model. However, we will learn in Chapter 6 that this is not true in general. For systems containing bimolecular reactions, the mean is also influenced by the fluctuations. In some systems, the mean of the stochastic model can be considerably larger than the deterministic prediction, and can lead to enhanced sensitivity of the network, known as "stochastic focusing" [121, 122].

Bistability: A bistable system has two stable steady states separated by an unstable steady state. In a deterministic framework, such a system settles to the steady state whose basin of attraction contains the initial condition. In a stochastic framework, however, the behavior is more complex: either steady state may be reached in different realizations regardless of the initial condition. This behavior is referred to as "stochastic switching" [58, 160], illustrated in Figure 1.6 for the Schlögl reaction, to be discussed in the following two chapters. The time-varying histogram, which was obtained from

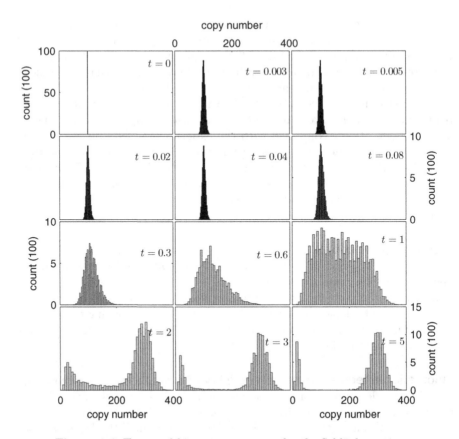

Figure 1.6 Temporal histogram progress for the Schlögl reaction.

10000 realizations, is unimodal initially and has a bimodal pattern at the end. This phenomenon is better visualized as the time-dependent 3-dimensional probability distribution shown in Figure 1.7. To study a bistable process (e.g., apoptosis, cell differentiation), single-cell technologies are necessary. Averaging over ensembles of cells, as done in a Western blot, does not allow one to distinguish between states. Using single-cell technologies, such as microscopy, a sample generated from a collection of cells under the same condition has proportions of cells in each state. The stochastic approach is necessary for capturing the variability in these experimental observations.

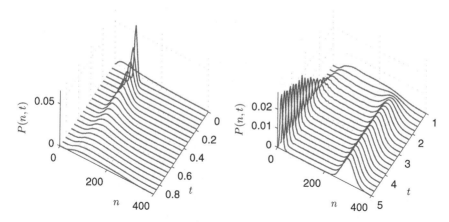

Figure 1.7 Temporal distribution progress for the Schlögl reaction. *Left*: monomodal during the time interval $0 < t < 1$, *Right*: turns bimodal during later time interval $1 < t < 5$.

1.5 Stochastic Approaches

The pioneering works of Kramers [84] and Delbrück [27] provided the impetus for the application of stochastic approaches to chemical kinetics. The most common formulation of stochastic models for biochemical networks is the chemical master equation (CME). The analytical nature of the early stochastic approaches made it highly complicated and, in some cases, intractable altogether. That is why early analytical stochastic approaches received little attention in the biochemical community. Later, the situation changed with the increasing computational power of modern computers. The ground-breaking work of Gillespie [51, 52] presented an algorithm for numerically generating sample trajectories of the abundances of chemical species in chemical reaction networks. The so-called "stochastic simulation algorithm," or "Gillespie algorithm," can easily be implemented in any programming or scripting language that has a pseudorandom number generator. Several software packages implementing the algorithm have been developed [15, 17, 49, 81, 129, 130]. For a survey of stochastic simulation approaches, the reader is referred to two recent reviews [113, 155]. Different stochastic approaches and their interrelationships are depicted in Figure 1.8.

For large biochemical systems, with many species and reactions, stochastic simulations (based on the original Gillespie algorithm) become computationally demanding. Recent years have seen a large interest in improving the efficiency/speed of stochastic simulations by modification/approximation of the original Gillespie algorithm. These improvements include the "next

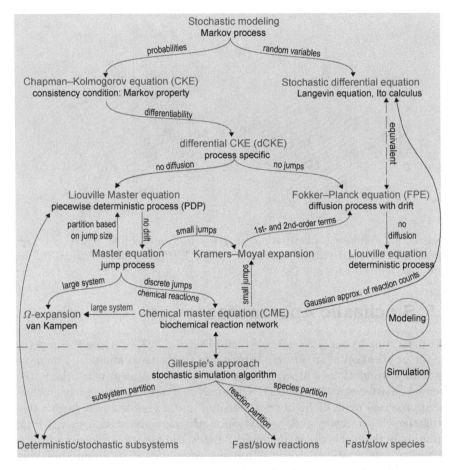

Figure 1.8 Classification of stochastic processes: Interrelationships for various stochastic approaches. Figure adopted from [158].

reaction" method of Gibson and Bruck [50], the "τ-leap" method [57] and its various improvements [20–22] and generalizations [23, 90] and the "maximal time step method" [124], which combines the next reaction and the τ-leap methods.

While stochastic simulations are a practical way to realize the CME, analytical approximations offer more insights into the influence of noise on cell function. Formally, the CME is a continuous-time discrete-state Markov process [52, 75, 143]. For gaining intuitive insight and a quick characterization of fluctuations in biochemical networks, the CME is usually approximated analytically in different ways [61, 75], including the frequently used chemical Langevin equation (CLE) [56, 76, 144, 174], the linear noise approximation (LNA) [34, 65, 139, 140], and the two-moment approximation (2MA) [44, 58, 62].

Of the analytical approaches mentioned above, we focus in Chapter 6 on the 2MA approach because of its representation of the coupling between the mean and (co)variance. The traditional Langevin approach is based on the assumption that the time-rate of abundance (copy number or concentration) or the flux of a component can be decomposed into a deterministic flux and a Langevin noise term, which is a Gaussian (white noise) process with zero mean and amplitude determined by the system dynamics. This separation of noise from the system dynamics may be a reasonable assumption for *external noise* that arises from the interaction of the system with other systems (such as the environment), but cannot be assumed for internal noise that arises from within the system [13, 30, 74, 117, 128, 141]. As categorically discussed in [76], internal noise is not something that can be isolated from the system because it results from the discrete nature of the underlying molecular events. Any noise term in the model must be derived from the system dynamics and cannot be presupposed in an ad hoc manner. However, the CLE does not suffer from the above criticism because Gillespie [56] derived it from the CME description. The CLE allows much faster simulations compared to the exact stochastic simulation algorithm (SSA) [52] and its variants. The CLE is a stochastic differential equation (dealing directly with random variables rather than moments) and has no direct way of representing the mean and (co)variance and the coupling between the two. That does not imply that CLE, like the LNA, which has the same mean as the solution of the deterministic model, ignores the coupling.

The merits of the 2MA compared to alternative approximations have been discussed in [58, 62, 149]. In [44], the 2MA is developed as an approximation of the master equation for a generic Markov process. In [58], the 2MA framework is developed under the name "mass fluctuation kinetics" for biochemical networks composed of elementary reactions. The authors demonstrate that the 2MA can reveal new behavior such as stochastic focusing and bistability. Another instance of the 2MA is proposed in [61, 62] under the names "mean-field approximation" and "statistical chemical kinetics." Again, the authors assume elementary reactions, so that the propensity function is at most quadratic in concentrations. The authors evaluate the accuracy of the 2MA against the alternatives (such as LNA) for a few toy models. The derivation of the 2MA for more general systems with nonelementary reactions is one motivation for our derivation in this book.

The 2MA approaches referred to above assume absolute concentrations (copy number divided by some fixed system-size parameter). In systems biology, however, models often use relative concentrations that have arbitrary units [26, 108, 109, 156]. In general, the concentration of each component in the system may have been obtained by a different scaling parameter, rather than using a global system size. For such models, the above-mentioned approaches need modification. This was another motivation for our derivation in this book. We develop a compact form of the 2MA equations—a system of ODEs for the dynamics of the mean and (co)variance of the continuous-time discrete-state Markov process that models a biochemical reaction system by the CME. This is an extension of previous derivations, taking into account arbitrary concentrations and nonelementary reactions. The compact form, obtained by careful selection of notation, for our derivation allows for an easy interpretation. Using these analytical results, we develop our 2MA model of the fission yeast cell cycle, which has two sets of ODEs: one set for the mean protein concentrations and the other set for concentration (co)variances. Numerical simulations of our model show a considerably different behavior. Especially, for the $wee1^-$ $cdc25\Delta$ mutant (hereinafter referred to simply as double mutant), the timings of S-phase and M-phase are visibly different from those obtained for a deterministic model because of the oscillatory behavior of the key regulator. Since the 2MA is only an approximation, we investigate its validity by comparing the statistics computed from the 2MA model with experimental data.

1.6 Outline of the Text

The remainder of the text is organized as follows.

Chapter 2: Representations of biochemical reactions are reviewed. We introduce key concepts such as concentration, system size, and reaction count, and provide a motivation for more complex reaction networks. The reaction rate is expressed in terms of the time derivative of reaction count, which is a more natural way than in terms of concentrations. Examples of chemical reactions have been chosen to illustrate certain key ideas. The standard modification (isomerization) is the simplest possible reversible reaction system. The complexity is gradually increased by heterodimerization (illustrating a reversible bimolecular reaction). Simple networks of more than one reaction are illustrated by the Lotka–Volterra model and the enzyme kinetic reaction system. Branching in a reaction network is illustrated by an example that also illustrates the phenomenon of stochastic focusing. Bistability, an important phenomenon, is illustrated by the Schlögl model. Finally, a biologically more relevant system, a simple gene regulatory network, is introduced to illustrate the key idea that noise matters even with large copy numbers because it propagates from transcription to translation. The Matlab code is provided for most examples in order to encourage the reader to play with the models.

Chapter 3: This chapter provides an informal discourse on the notion of randomness and its mathematical representation by random variables. Working with simple reaction examples, key concepts in probability, random variables, and stochastic processes are introduced along the road. The relevant terms are introduced when needed, in contrast to the usual approach of first giving definitions without setting the stage. Probability theory is kept to a minimum and the focus is more on convincing the reader through intuition.

Chapter 4: Key concepts in probability, random variables, and stochastic processes are reviewed. This chapter cannot serve as a complete introduction to probability theory but should be enough for those who have been introduced to probability and random variables at some stage in their academic career. To keep things short, only those probability distributions that are important for later chapters have been included. These include the exponential distribution (vital for Markov processes), the Poisson distribution (underlies the famous Poisson process), and the uniform distribution (important for random number generation). A first flavor of stochastic processes is given by presenting a detailed account of the Poisson process.

Chapter 5: The stochastic framework for modeling subcellular biochemical systems is presented. In particular, an effort is made to show how the notion of propensity, the chemical master equation, and the stochastic simulation algorithm arise as consequences of the Markov property. This connection is not obvious from the relevant literature in systems biology. Moreover, we review various analytical approximations of the chemical master equation. The notation has been carefully chosen to make it easy for the reader to see how different approximations are related to each other. Examples introduced in Chapter 2 are revisited in a stochastic setting. For each example, simulation results are presented.

Chapter 6: This chapter develops a compact form of the 2MA equations—a system of ODEs for the dynamics of the mean and (co)variance of the continuous-time discrete-state Markov process that models a biochemical reaction system by the CME. This is an extension of previous derivations, taking into account relative concentrations and nonelementary reactions. The compact form, obtained by careful selection of notation, allows for an easy interpretation.

Chapter 7: This chapter takes the Tyson–Novák model for the fission yeast cell cycle as a case study. This deterministic model is a practical example using nonelementary reactions and relative concentrations, the two central features of our extended 2MA approach. This will allow us to investigate the price of higher-order truncations by comparing the simulated cycle time statistics with experiments.

Chapter 8: This chapter deals with general Markov processes that can have both continuous and jump character. A general system of differential equations is derived for the general Markov process. Then it is illustrated how different processes can arise as special cases of the general process. That leads to a family tree of stochastic models that is sketched in a detailed diagram.

Chapter 9: In this chapter we review selected publications on noise and stochastic modeling, including those linked to experimental studies. Due to the wide range of experimental technologies used to generate data, and because the importance of this to the analysis, we cannot reproduce those studies in a book like this. The selection of a few papers is to demonstrate the relevance of noise and stochastic modeling to state-of-the-art molecular and cell biology.

Chapter 2

Biochemical Reaction Networks

This chapter is a basic introduction to chemical reactions and chemical species. Different ways of quantifying the abundance of molecules lead to the notions of concentration. Similarly, a deterministic quantification of how fast a reaction proceeds in time leads to notions such as reaction rate and rate constant. Representation of biochemical reaction schemes is reviewed. Deterministic description of reaction networks in terms of reaction rates is described.

2.1 The Notion of a Chemical Reaction

Molecules inside the cell undergo various transformations. For example, a molecule can transform from one kind to another, two molecules of the same or different kinds can combine to form another molecule of a third kind, and so on. At the basic level these transformations are known as chemical reactions. In the context of chemical reactions, a molecule is said to be (an instance) of a certain species. Similarly, a chemical reaction is said to be (an instance) of a certain channel. The chemical species are denoted by roman uppercase letters. A single molecule of a species A is referred to as an A-molecule. The chemical reaction is written schematically as an arrow with reactants on the left and products on the right. Thus an A-molecule could transform to a B-molecule:

$$A \to B,$$

a conversion or modification or isomerization. An A-molecule could associate with a B-molecule to form a non-covalently-bound complex:

$$A + B \to C,$$

an *association* or synthesis. The complex C-molecule could dissociate into an A- and a B-molecule:

$$C \to A + B,$$

a dissociation or decomposition. A species that is not of interest to us (e.g., because its abundance does not change over time) is represented by the symbol

∅ and referred to as the "null species." So the reaction

$$A \rightarrow \varnothing$$

represents the degradation of an A-molecule to a form not of interest to us. Similarly, the production of a B-molecule is written as

$$\varnothing \rightarrow B$$

when the reactants are disregarded. These reactions are said to be *elementary* and *irreversible*; elementary in the sense that each one takes one basic step (association, dissociation, conversion) to complete and irreversible because the change is only in one direction. They never exist in isolation, but always in combination with each other. So, what we usually describe as a chemical reaction can always be broken down into a mechanism that consists of combinations of these three elementary processes. For example, the probable mechanism of the chemical reaction

$$A + B \rightleftharpoons C$$

would be

$$A + B \rightleftharpoons AB \rightarrow C,$$

where C is a covalent modification of AB. Each half ($\rightharpoonup$ or $\leftharpoonup$) of the double arrow ($\rightleftharpoons$) denotes one of the elementary reactions. Thermodynamically, all chemical reactions are reversible and consist of a forward reaction and a reverse reaction. Thus when we write an irreversible reaction, it will either represent the forward or backward step of a reversible reaction, or a simplification (i.e., approximation) of a reversible reaction by an irreversible one.

2.2 Networks of Reactions

Imagine molecules of s chemical species homogeneously distributed in a compartment of constant volume V at thermal equilibrium and interacting through r irreversible reaction channels. A reaction channel either is elementary or may represent a simplification of multiple elementary steps into a single step. Any reversible (bidirectional) reaction can be listed as two irreversible reactions. We symbolize the ith species with X_i and the jth reaction channel with R_j. The abundance of X_i present in the system at time t can be described by the copy number $N_i(t)$. The total copy number n^{tot} of all species indicates how large the system is. Since a large/small value of n^{tot} usually implies a large/small volume, the volume V can also indicate the size of the system. Any such parameter can be used as the *system size* and is usually denoted by

Ω. The copy number is usually divided by the system size, and the quantity thus obtained,

$$X_i(t) = \frac{N_i(t)}{\Omega},$$

is referred to as the concentration. The choice of the system size Ω depends on the kind of concentration one would like to define.

Molar Concentrations: For molar concentrations, in units $M \equiv \text{mol}/\text{L}$, the system size is chosen as $\Omega = N_A V$, where Avogadro's constant

$$N_A = 6.022 \times 10^{23} \, \text{mol}^{-1}$$

(correct to four significant digits) is the number of molecules (or any elementary entities) in one mole. If the volume is given in liters (L) and concentration in molar (M), then the unit of system size Ω is $\text{mol}^{-1} \times \text{L} = M^{-1}$. The molar unit (M) is too large for very small concentrations, which are better specified in smaller units including nanomolar (nM), micromolar (μM), and millimolar (mM). Suppose the proteins in a cell of volume $V = 30 \, \text{fL}$ are measured in nanomolar $(\text{nM})^{-1}$; then the computation of the system size proceeds like this:

$$\Omega = N_A V = \left(6.022 \times 10^{14} \, (\text{n mol})^{-1} \right) \times \left(3 \times 10^{-14} \, \text{L} \right) \approx 18 \, (\text{nM})^{-1} .$$

Sometimes, the volume is chosen so that $\Omega = 1 \, (\text{nM})^{-1}$ for the resulting convenience that each nanomolar concentration is numerically equal to the corresponding copy number.

Relative concentrations: For relative concentrations, the system size is chosen to give dimensionless concentrations. One simpler way to obtain relative concentrations is by choosing $\Omega = n^{\text{tot}}$, so that each concentration is just a fraction of two copy numbers. Take the isomerization reaction as an example whereby proteins are converted back and forth between the unmodified form U and the modified form W such that the total number n^{tot} of protein molecules remains constant. The relative concentrations in this example are the fractions

$$X_U(t) = \frac{N_U(t)}{n^{\text{tot}}} \quad \text{and} \quad X_W(t) = \frac{n^{\text{tot}} - N_U(t)}{n^{\text{tot}}}$$

of proteins in the inactive and active form, respectively. For some systems it is more appropriate to introduce a different scaling parameter Ω_i for each component i if the copy numbers N_i differ in magnitude to keep X_i of the same order $\mathcal{O}(1)$. That can be obtained by defining relative concentration as

$$X_i = \frac{N_i}{C_i \Omega},$$

that is, the concentration N_i/Ω divided by a characteristic concentration C_i. In that case, each scaling parameter can be expressed as $\Omega_i = C_i \Omega$. This will be of concern to us in the following chapter. In this chapter, we stick to the simpler case.

The reaction channel R_j will be represented by the general scheme

$$\underline{S}_{1j}X_1 + \cdots + \underline{S}_{sj}X_s \xrightarrow{k_j} \bar{S}_{1j}X_1 + \cdots + \bar{S}_{sj}X_s. \tag{2.1}$$

The participation of individual species in the reaction is indicated by *stoichiometries*, or *stoichiometric coefficients*, written beside them. Thus, the coefficient $\underline{S}_{ij}$ (on the left) represents the participation of X_i as a reactant and $\bar{S}_{ij}$ (on the right) is the corresponding participation as a product. The rate constant, or coefficient, k_j, written over the reaction arrow informs us about the assumed reaction kinetics, and will be explained later. The coefficient will be omitted when we do not want to attach any assumed reaction kinetics to the above reaction scheme. The progress of channel R_j is quantified in this text by the *reaction count* $Z_j(t)$, defined as the number of occurrences of R_j during the time interval $[0, t]$. One occurrence of R_j changes the copy number of X_i by $S_{ij} = \bar{S}_{ij} - \underline{S}_{ij}$, the (i, j)th element of the *stoichiometry matrix* S. During the time interval $[0, t]$, the change in the copy number of X_i contributed by R_j is thus $S_{ij}Z_j(t)$. The total change in the copy number is the sum of contributions from all reactions:

$$N_i(t) = N_i(0) + \sum_{j=1}^{r} S_{ij} Z_j(t). \tag{2.2}$$

Thus changes in copy numbers are determined by stoichiometries and reaction counts. We need to caution the reader against a potential confusion between the term *reaction count* and a similar term *reaction extent*. Since the copy numbers appearing in the above equation are in units of molecules, we can also interpret the reaction count $Z_j(t)$ as number of molecules of a hypothetical substance in terms of which the other copy numbers are expressed. Dividing the above equation by N_A will change the measurements from molecules to moles, and the reaction count is replaced by the reaction extent $Z_j(t)/N_A$. Following the usual vector notation, we write $N(t)$ for the s-vector of copy numbers, $X(t)$ for the s-vector of concentrations, and $Z(t)$ for the r-vector of reaction counts. The above conservation relation can be written in vector notation:

$$N(t) = N(0) + S\,Z(t)\,. \tag{2.3}$$

Dividing by Ω gives the corresponding relation in concentrations:

$$X(t) = X(0) + \frac{S\,Z(t)}{\Omega}\,. \tag{2.4}$$

The quantity $Z_j(t)/\Omega$ is referred to as the *degree of advancement* of the reaction channel and replaces the role of reaction count in converting the progress of reaction to species concentration.

The copy number $N(t)$, the concentration $X(t)$, and the reaction count $Z(t)$ are alternative ways to describe our system. Description in terms of these *macroscopic variables* is done in the hope that they approximately satisfy an autonomous set of deterministic (differential or difference) equations. Because of the ease of analysis, differential equations are always preferred over the difference equation. However, the reactions are discrete events in time, which means that the copy numbers do not vary continuously with time. That would require the adoption of difference equations. The situation is made even more complicated by two problems. Firstly, the occurrence time of a reaction is a random quantity because it is determined by a large number of microscopic factors (e.g., positions and momenta of the molecules involved). The second problem arises when more than one type of reaction can occur. The type of reaction to occur is also a random quantity for the same reasons mentioned above. Therefore, the deterministic description needs a few simplifying assumptions. Alternatively, the macroscopic variables are formulated as stochastic processes. Such a stochastic description in terms of macroscopic variables is *mesoscopic*.

Throughout this text, we will use a couple of academic examples. They are chosen to demonstrate different ideas and methods in the discussion. For further examples of simple biochemical networks and a discussion of their relevance to molecular and cell biology, the reader is referred to [157].

Example 2.1 (Standard modification) Consider a protein that can exist in two different conformations, or forms, an *unmodified* form U and a *modified* form W. The protein changes between the two forms by the reversible isomerization reaction

$$U \underset{k_u}{\overset{k_w}{\rightleftharpoons}} W \tag{2.5}$$

composed of a modification (forward) channel with rate constant k_u and a demodification (reverse) channel with rate constant k_w. The reaction scheme (2.5) also represents the opening and closing of an ion channel and similar systems with two-state conformational change. Since the two reactions are not influenced by any external catalyst (e.g., an enzyme), the scheme (2.5) will be referred to as the *standard modification*. This example was used in the introductory chapter to illustrate ideas of identifiability and species extinction or depletion.

Example 2.2 (Heterodimerization) Consider the reversible heterodimerization

$$X_1 + X_2 \underset{k_2}{\overset{k_1}{\rightleftharpoons}} X_3 \,. \tag{2.6}$$

Here the forward reaction is the association of a receptor X_1 and a ligand X_2 to form a heterodimer (complex) X_3. The backward reaction is the dissociation of the heterodimer back into the two monomers. The parameters k_1 and k_2 are the respective association and dissociation rate constants. This example is the simplest one with a bimolecular reaction.

Example 2.3 (Lotka–Volterra model) Consider the process whereby a reactant A, replenished at a constant rate, is converted into a product B that is removed at a constant rate. The reaction will reach a steady state but cannot reach a chemical equilibrium. Suppose the process can be decomposed into three elementary steps:

$$X_1 + A \longrightarrow 2X_1,$$
$$X_1 + X_2 \longrightarrow 2X_2,$$
$$X_2 \longrightarrow B \,.$$

The first two reactions are examples of an autocatalytic reaction: the first one is catalyzed by the reactant X_1 and the second by the reactant X_2. This simple reaction scheme was proposed as a simple mechanism of oscillating reactions [67, 94]. Although the scheme illustrates how oscillation may occur, known oscillating chemical reactions have mechanisms different from the above. For an in-depth treatment of biochemical oscillations, the reader is referred to [42, 59, 106, Chapter 9]. This type of process is found in fields other than chemistry; they were investigated in the context of population biology by Lotka [94] and Volterra [164]. Due to the frequent appearance of this latter context in the literature, we rewrite the above reaction scheme as

$$\left. \begin{aligned} X_1 + A &\xrightarrow{k_1} 2X_1, \\ X_1 + X_2 &\xrightarrow{k_2} 2X_2, \\ X_2 &\xrightarrow{k_3} \varnothing, \end{aligned} \right\} \tag{2.7}$$

as a system of two interacting species: X_1 (the prey) and X_2 (the predator). The food (substrate) A is available for X_1, which reproduces, with rate

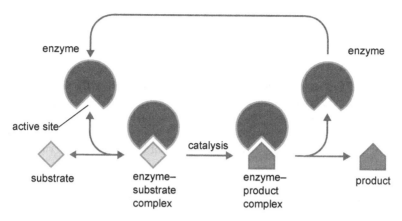

Figure 2.1 Enzyme-catalyzed conversion of a substrate to a product. The enzyme binds to the substrate to make its conversion to product energetically favorable. Figure based on an illustration in Alberts et al. [4].

coefficient k_1, after consuming one unit of A. An encounter between the two species, with rate coefficient k_2, results in the disappearance of X_1 and the replication of X_2. This is the only way X_1 dies (degrades), whereas X_2 has a natural death (degradation) with rate coefficient k_3. The food A is assumed to be constantly replenished so that the copy number n_A remains constant. This example serves the purpose of a simple system containing a bimolecular reaction and the resulting influence of (co)variance on the mean (Chapter 6).

Example 2.4 (Enzyme kinetic reaction) In biological systems, the conversion of a substrate to a product may not be a thermodynamically feasible reaction. However, specialized proteins called enzymes ease the job by binding to the substrate and lowering the activation energy required for conversion to the product, as depicted in Figure 2.1. Represented in reaction notation,

$$E + S \longrightarrow E + P,$$

the enzymatic reaction is thought to be accomplished in three elementary steps:

$$E + S \underset{k_2}{\overset{k_1}{\rightleftharpoons}} ES \overset{k_3}{\longrightarrow} E + P. \tag{2.8}$$

Here the enzyme E catalyzes a substrate S into a product P that involves an intermediary complex ES. Note that we have not placed any rate constant over the arrow in the original reaction because we do not specify any assumed kinetics in that notation. Later we will learn that it is possible to approximate the three elementary reactions by a single reaction,

$$S \xrightarrow{k_{\text{eff}}} P,$$

with an effective rate coefficient k_{eff} that represents the assumed approximate kinetics. Intuitively, k_{eff} will be a function of the enzyme abundance. We include this example because this type of reaction appears frequently in the literature. It also serves the purpose of a simple system containing a bimolecular reaction and allows demonstration of how mass conservation leads to a simplified model.

Example 2.5 (Schlögl model) The Schlögl model is an autocatalytic, trimolecular reaction scheme, first proposed by Schlögl [137]:

$$A + 2X \xrightleftharpoons[k_2]{k_1} 3X, \quad B \xrightleftharpoons[k_4]{k_3} X. \tag{2.9}$$

Here the concentrations of A and B are kept constant (buffered). This example, mentioned in the introduction, serves to illustrate the need for a stochastic approach to model systems with bistability and the associated behavior known as "stochastic switching."

Example 2.6 (Stochastic focusing) This example was first described in [121] to demonstrate a behavior phenomenon known as "stochastic focusing." The branched reaction network comprised the following reaction channels:

$$\varnothing \xrightleftharpoons[k_d]{k_s} S, \quad \varnothing \xrightleftharpoons[k_a X_S]{k_i} I \xrightarrow{k_p} P \xrightarrow{1} \varnothing. \tag{2.10}$$

Here the product P results from the irreversible isomerization of its precursor I, an intermediary chemical species. This isomerization is inhibited by a signaling chemical species S that is synthesized and degraded by independent mechanisms.

Example 2.7 (Gene regulation) As pointed out earlier, oscillating chemical reactions have mechanisms different from the simple and intuitive Lotka–Volterra scheme. Those familiar with dynamical systems theory will recall that such a kinetic system can oscillate only if both activation and inhibition are present in the form of a feedback loop. Such feedback loops exist in gene expression, where the protein product serves as a transcription factor and represses transcription. A simplified regulatory mechanism is illustrated in Figure 2.2. The protein product from gene expression binds to a regulatory region on the DNA and represses transcription. The regulatory mechanism

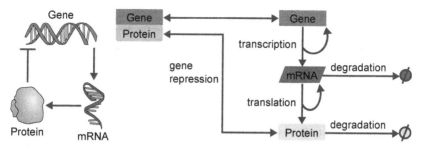

Figure 2.2 Gene regulation: a simplified model. *Left*: cartoon representation. *Right*: reaction pathways.

is simplified by not showing the contributions of RNA polymerase and any cofactors. The reaction scheme for the system is

$$
\left.
\begin{aligned}
G &\xrightarrow{\ k_m\ } G + M && \text{(transcription)}, \\
M &\xrightarrow{\ k_p\ } M + P && \text{(translation)}, \\
G + P &\underset{k_u}{\overset{k_b}{\rightleftharpoons}} GP && \text{(binding/unbinding)}, \\
M &\xrightarrow{k_m^-} \varnothing, \ P \xrightarrow{k_p^-} \varnothing && \text{(degradation)},
\end{aligned}
\right\}
\tag{2.11}
$$

where the gene G is transcribed to the mRNA M with rate constant k_m, the mRNA is translated to the protein P with rate constant k_p, and the protein binds to (and represses) the gene with rate constant k_b and unbinds back with rate constant k_u. The mRNA and protein are degraded with respective rate constants k_m^- and k_p^-.

Synthetic gene regulation: The above idea of a simple feedback loop has motivated several researchers to construct such feedback transcriptional regulatory networks in living cells [11]. These investigators found that an increased delay in the feedback loop increases the dynamic complexity of the synthetic transcription system. A feedback loop with one repressor protein constructed by Becskei and Serrano [12] exhibited on and off transitions. Another loop with two repressor proteins, constructed by Gardner, Cantor, and Collins [48], manifested bistability in the on and off states. Yet another loop with three repressor proteins, constructed and termed "repressilator" by Elowitz and Leibler [40], exhibited oscillations.

Repressilator: The repressilator is a milestone of synthetic biology because its shows that gene regulatory networks can be designed and implemented to

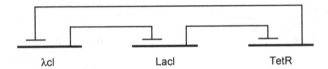

Figure 2.3 The repressilator gene regulatory network.

perform a novel desired function [40]. The repressilator consists of three genes, λcl, Lacl, and TelR, connected in a feedback loop. As depicted in Figure 2.3, each gene product represses the next gene in the loop, and is repressed by the previous gene. In addition, not shown in the figure, green fluorescent protein is used as a reporter so that the behavior of the network can be observed using fluorescence microscopy.

Cell cycle: The cycle through which cells grow and duplicate their DNA before eventually dividing into two daughter cells is of central importance to the realization of higher levels of biological organization. Underlying the cell cycle and its regulation are complex mechanisms, realized through large reaction networks. Due to its complexity, the cell cycle is investigated as a case study in Chapter 7.

2.3 Deterministic Description

Suppose that reactions occur so frequently that the reaction count $Z(t)$ can be approximated by a continuous quantity $z(t)$. This assumption requires that a large number of reactant molecules be freely available (no crowding) in a large volume so that they can react easily. It also requires that the energy and orientation of reactant molecules favor the reaction, a fact summarized in a rate constant. Large numbers of molecules also mean that a change resulting from a single occurrence of a reaction is relatively small. This means that the copy number $N(t)$ can be approximated by a continuous quantity $n(t)$. The concentration $X(t)$ is similarly approximated by a continuous quantity $x(t)$. In a deterministic description, equations (2.3) and (2.4) respectively translate to

$$n(t) = n(0) + S\,z(t) \tag{2.12}$$

and

$$x(t) = x(0) + \frac{S\,z(t)}{\Omega}. \tag{2.13}$$

Taking the time derivatives gives us the net chemical fluxes:

$$\dot{n}(t) = S\,\dot{z}(t), \quad \dot{x}(t) = S\,\frac{\dot{z}(t)}{\Omega}. \tag{2.14}$$

Here the time derivative $\dot{x}$ is the net concentration flux and $\dot{n}$ is the net copy-number flux. Note that our usage of the term "chemical flux" differs from IUPAC[1], which defines it in terms of moles. The above equations are useful only if a relationship between the time derivative on the right and the abundance variable ($n(t)$ or $x(t)$) is established. Suppose a relation can be mathematically represented as

$$\dot{z} = \hat{v}(n) = \Omega v(x), \tag{2.15}$$

where the vectors $\hat{v}(x)$ and $v(x)$ are referred to here as the *conversion rate* and the *reaction rate*, respectively. The conversion rate is here defined as reaction count per unit time, a slight difference with the standard definition in [1] as the time derivative $\dot{z}/N_A$ of the extent of reaction. The reaction rate is defined as reaction count per unit time divided by the system size. The notation $v\left(x(t)\right)$ is based on the assumption that the reaction rate depends only on the concentrations of the reactants. This is a realistic assumption in many reactions at constant temperature. In general, the reaction rate can depend on temperature, pressure, and the concentrations or partial pressures of the substances in the system.

The functional form $v_j(\cdot)$ of the rate of R_j is called the *rate law* (or kinetic law), which is a result of the modeling assumptions about the particular reaction channels. It is only after specifying a rate law that the above ODEs can characterize a particular biochemical reaction network. Without that specification, the above ODEs only represent a consistency condition imposed by mass (or substance) conservation of reactants and products. Incorporating the rate law specification (2.15) into the ODEs (2.14) leads to the deterministic *chemical kinetic equations*

$$\dot{n}(t) = S\,\hat{v}\left(n(t)\right), \quad \dot{x}(t) = S\,v\left(x(t)\right). \tag{2.16}$$

There is a large class of chemical reactions in which the reaction rate is proportional to the concentration of each reactant raised to some power:[1]

$$v_j(x) = k_j \prod_{i=1}^{s} x_i^{g_{ij}}, \quad \hat{v}_j(n) = \hat{k}_j \prod_{i=1}^{s} n_i^{g_{ij}}, \tag{2.17}$$

which is called *a rate law with definite orders* [102]. The rate constant k_j summarizes factors such as activation energy and proper orientation of the reactant molecules for an encounter leading to the reaction. The rate constant k_j can be interpreted as the factor of the reaction rate that does not

---

[1] Since 0^0 is undefined, the product $\prod_{i=1}^{s}$ must exclude i for which both x_i and g_{ij} are zero.

depend on reactant concentrations. The conversion rate constant $\hat{k}_j$ has a
similar interpretation as the factor of the extensive reaction rate that does
not depend on the reactant copy numbers. Recall that while the units of
$\hat{k}_j$ are always $\sec^{-1}$, the units of k_j additionally depend on the units used
for the concentration x. The exponent g_{ij} is the order with respect to the
species X_i. The sum of orders for a particular reaction channel is the overall
order. For elementary reactions, the orders g_{ij} are the same as the reactant
stoichiometries $\underline{S}_{ij}$:

$$v_j(x) = k_j \prod_{i=1}^{s} x_i^{\underline{S}_{ij}}, \quad \hat{v}_j(n) = \hat{k}_j \prod_{i=1}^{s} n_i^{\underline{S}_{ij}} . \tag{2.18}$$

This rate law is called *mass-action kinetics* [66] and is justified by collision
theory and transition state theory [71, 102, 171]. The mass-action kinetics
should not be confused with the closely related law of mass action, which
is obtained by equating the forward and backward reaction rates (according
to the above rate law) of a reversible reaction. Reactions that cannot be
described by rate laws like (2.17) are said to have *no definite order*. For
such a reaction, the rate law depends on the assumptions involved in the
approximation of the constituent reaction channels. Examples of such rate
laws include Michaelis–Menten kinetics, Hill kinetics, and competitive inhibition [25, 43, 66]. A family tree of deterministic ODE models is sketched in
Figure 2.4. The ODEs in their most general form are rarely used in systems
biology. Equation (2.16) is the most common representation to describe the
continuous changes in concentration $x(t)$ in terms of the network structure,
encoded by the stoichiometry matrix S, and the network kinetics, encoded by
the rate law $v(\cdot)$. Note that the kinetic parameters such as the rate constant
k and the kinetic order g are incorporated in the rate law. Further variations
emerge through an implicit assumption about the underlying biophysical
environment in which reactions take place. Assuming basic mass-action-type
kinetics, the kinetic order g_{ij} of the rate law will typically take the value 1
or 2 (dimerization). Further quasi-steady-state assumptions for intermediate
complexes can simplify into Michaelis–Menten type kinetic models. The left
branch allows for noninteger kinetic orders and takes two routes that depend
on the semantics [161]. Simplified power-law models (e.g., S-Systems [163])
assume very little knowledge about the biophysical structure of the environment in which reactions take place. These models distinguish between positive
and negative contributions (pos/neg kinetic orders) and different strengths
of activation/inhibition. On the other hand, criticizing the assumption of a
homogeneous and well mixed environment (underlying the right branch) leads
to noninteger (but positive) kinetic orders. A detailed kinetic power-law model
would thus arguably represent the biophysical environment more accurately

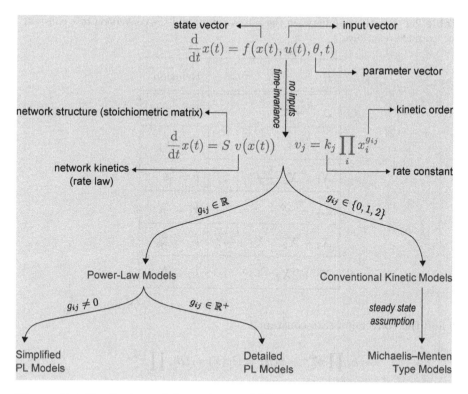

Figure 2.4 Family tree of deterministic ODE models. For chemical reaction networks, the general ODE formulation simplifies to a decomposition into the stoichiometry matrix (encoding the network structure) and the rate law (encoding the network kinetics). A large class of chemical reactions have a rate law with definite (kinetic) orders, of the form (2.17). Restricting and broadening the range of values of the kinetic order g_{ij} allows further classification.

than the conventional mass-action model. On the other hand the simplified power-law model admits a more phenomenological interpretation. A drawback of the power-law models is that of additional parameters, the kinetic orders, they introduce. The more parameters a model has, the more difficult it is to identify a unique set of parameter values from experimental time-course data.

Relationship between k and $\hat{k}$: We can combine the defining relationship (2.15) with the rate law (2.17) to get a relationship between the rate constant

Table 2.1 Relationship between the rate constant and the conversion rate constant for example reactions.

R_j	Relation
$\varnothing \xrightarrow{k_j} X$	$\hat{k}_j = \Omega k_j$
$X \xrightarrow{k_j} ?$	$\hat{k}_j = k_j$
$X_1 + X_2 \xrightarrow{k_j} ?$	$v = \frac{k_j}{\Omega}$
$2X \xrightarrow{k_j} ?$	$\hat{k}_j = \frac{k_j}{\Omega}$
$X_1 + X_2 + X_3 \xrightarrow{k_j} ?$	$\hat{k}_j = \frac{k_j}{\Omega^2}$
$X_1 + 2X_2 \xrightarrow{k_j} ?$	$\hat{k}_j = \frac{k_j}{\Omega^2}$

k and the conversion rate constant $\hat{k}$:

$$\hat{k}_j \prod_{i=1}^{s} n_i^{g_{ij}} = \hat{v}(n) = \Omega v(x) = \Omega k_j \prod_{i=1}^{s} x_i^{g_{ij}}.$$

Now invoke the defining relationship $n = \Omega x$ to obtain

$$\hat{k}_j = \frac{k_j}{\Omega^{K_j - 1}}, \tag{2.19}$$

where $K_j = \sum_{i=1}^{s} g_{ij}$, which, for elementary reactions, is simply $K_j = \sum_{i=1}^{s} \underline{S}_{ij}$. The relationship for sample elementary reactions is illustrated in Table 2.1. The table suggests that the two types of rate constants are equal for monomolecular reactions.

Matlab implementation: To implement rate laws of the form (2.17) in Matlab [96], the standard Matlab data type *function handle* can be employed. We will need Matlab representations of our mathematical quantities. Let us collect the species concentrations x_i (at a certain time) in an $s \times 1$ column vector x, the reaction rate constants k_j in an $r \times 1$ column vector k, and the exponents g_{ij} (which equal $\underline{S}_{ij}$ for mass-action kinetics) of the rate law (2.17) in an $s \times r$ matrix G. Then the Matlab representation v of the rate law $v(\cdot)$ defined elementwise in (2.17) takes the following form:

M-code 2.1 makeRateLaw: implements rate law with definite orders (2.17).

```
function v = makeRateLaw(k,G)
r = size(G,2);
i0 = (G==0);
i = ~i0 & (G~=1);
v = @RateLaw;
    function vofx = RateLaw(x)
        X = repmat(x,1,r);
        X(i0) = 1;
        X(i) = X(i2.^G(i);
        vofx = k.*prod(X)';
    end
end
```

```
v = @(x) k.*prod(repmat(x,1,r).^G)';
```

where r is the Matlab representation of the number r of reaction channels. Here the function handle v stores the mathematical expression following @(x). The standard Matlab notations .* and .^ represent the elementwise operations multiplication and exponentiation. The compact code above may not be efficient in dealing with a large network of many species and reactions. Specifically, the exponentiation and multiplication are computationally demanding. To avoid these unnecessary computations, the code is replaced by Matlab function makeRateLaw in M-code 2.1. Here the output v returned by the main function makeRateLaw is a function handle to the *nested function* RateLaw. Note how exponentiation is avoided for the obvious cases $g_{ij} = 0$ and $g_{ij} = 1$. In general, a rate law may not be expressible in the form (2.17) and has to be written on a case-by-case basis. Once such function (or handle) has been written for the rate law, a Matlab representation of the chemical kinetic equations (2.16) can be written and numerically solved with the following piece of Matlab code:

```
dxdt = @(t,x) S*v(x); % concentration ODE
[tout,xout] = ode15s(dxdt, [0 tf], x0); % solution
```

Here x0 is a column vector of initial concentrations and tf is the final (stop) time of the simulation. The solver ode15s returns the column vector tout of time points and the solution array xout with a row of concentrations for each time point.

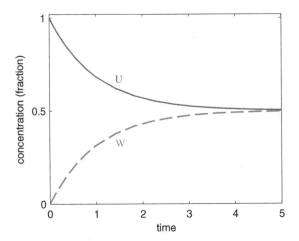

Figure 2.5 Time course of concentrations in the standard modification (2.20). Initially all molecules are assumed to be unmodified (U). The ordinate is the fraction of molecules in (un)modified form. Equilibrium is reached when the two fractions are equal. Both the rate constants were taken as $2\,\mathrm{sec}^{-1}$.

Example 2.8 (Standard modification) Consider the (de)modification of a protein between two forms by the reaction scheme (2.5). Suppose there are n^{tot} copies of this protein in a container, $n(t)$ of them being unmodified (in form U) at time t. The two reaction channels progress at the following conversion rates (listed on the right)

$$
\begin{array}{ll}
\mathrm{U} \xrightarrow{\;k_w\;} \mathrm{W} & \quad \hat{v}_w = k_w n \\[2mm]
\mathrm{W} \xrightarrow{\;k_u\;} \mathrm{U} & \quad \hat{v}_u = \left(n^{\mathrm{tot}} - n\right) k_u
\end{array}
\tag{2.20}
$$

and their difference gives the rate equation

$$
\dot{n} = -\hat{v}_w + \hat{v}_u = k_u n^{\mathrm{tot}} - (k_w + k_u)n \,.
$$

The rate equation for the unmodified fraction $x = n/n^{\mathrm{tot}}$ of all proteins is then

$$
\dot{x} = k_u - (k_w + k_u)x \,.
\tag{2.21}
$$

The Matlab implementation of this differential equation and its numerical solution will look like the following piece of code:

```
k = [2;2]; % rate constants
dxdt = @(t,n) k(2)-(k(1)+k(2))*x; % ODE
x0 = 1; % initial condition
```

```
[tout,xout] = ode15s(dxdt, [0 tf], x0); % solution
```

with the understanding that the Matlab workspace has values of variables k, tf, and x0, which correspond respectively to the rate constant $k = [k_w, k_u]$, the simulation stop time, and the initial fraction x^{init}. A typical time course is plotted in (2.21) wherein the fractions of molecules in the two forms are plotted against time. The above Matlab code can be rewritten in a way that lends itself to automatic code-writing. Toward that end, we write down the stoichiometry matrix S and the reaction rate vector v for this example:

$$S = \begin{bmatrix} -1 & 1 \end{bmatrix}, \quad v = \begin{bmatrix} v_w \\ v_u \end{bmatrix} = \begin{bmatrix} k_w x \\ (1-x)k_u \end{bmatrix}.$$

With these two quantities available, the above Matlab code can be replaced by

```
S = [-1 1]; % stoichiometry matrix
k = [2;2]; % rate constants
v = @(x) [k(2)*x; (1-x)*k(1)]; % reaction rate
dxdt = @(t,x) S*v(x)'; % rate equation
x0 = 1; % initial condition
[tout,xout] = ode15s(dxdt, [0 tf], x0); % solution
```

Here the first line assigns values to (the array) S, which corresponds to the stoichiometry matrix S. The second line assigns an expression to the function handle v, which corresponds to the rate law $v(\cdot)$. The next line defines the function handle dndt to represent the system of ODEs in question. The last line calls an ODE solver to solve the problem and returns the output arrays tout of time points and xout of concentration values. It can be seen from the above Matlab code that all we need is a representation S (a Matlab matrix) of the stoichiometry matrix S and a representation v (a Matlab function handle) of the reaction rate law $v(\cdot)$.

For the remainder of the text, we will mostly specify such quantities with an understanding that the reader can translate that information into the corresponding Matlab code.

Chemical equilibrium: When the modification rate v_w (in the last example) is balanced by the demodification rate v_u, chemical equilibrium is said to have occurred. In other words, the reversible reaction equilibrates or reaches the steady state. The steady-state fraction x^{ss} is the value of x that makes the time derivative in (2.21) zero, that is,

$$x^{\mathrm{ss}} = \frac{k_u}{k_w + k_u} .$$

Thus, in the steady state, a fraction $P_U = k_u/(k_u+k_w)$ of proteins are in the unmodified form and a fraction $P_W = k_w/(k_u+k_w)$ of them in the modified form. We can also say that a protein spends, on average, a fraction P_W of time in the modified form and a fraction P_U of time in the unmodified form. This interpretation proves very useful in reducing complicated reactions to single steps. Suppose the W form participates in another reaction $\mathrm{W} \xrightarrow{k_b} \mathrm{B}$ that occurs on a much slower time scale than two-state conformational changes between U and W. The overall complicated reaction

$$\mathrm{U} \underset{k_u}{\overset{k_w}{\rightleftharpoons}} \mathrm{W} \xrightarrow{k_b} \mathrm{B}$$

can be reduced to a single step $\varnothing \xrightarrow{k_b P_W} \mathrm{B}$ under the fast equilibration assumption for the reversible reaction.

Example 2.9 (Heterodimerization) Recall the reversible heterodimerization depicted in the reaction scheme (2.6). Let $x_1(t)$, $x_2(t)$, and $x_3(t)$ denote the respective time-dependent molar concentrations of receptor X_1, ligand X_2, and heterodimer X_3. The reaction network has to satisfy two conservation relations:

$$x_1 + x_3 = q_1, \quad x_2 + x_3 = q_2, \tag{2.22}$$

where q_1 and q_2 are constants determined by the initial conditions. Using these to express x_1 and x_2 in terms of x_3, the system state can be represented by tracking only species X_3. The reaction rates according to the mass-action kinetics follow from (2.18) to be (each listed to the right of the corresponding reaction channel)

$$\mathrm{X}_1 + \mathrm{X}_2 \xrightarrow{k_1} \mathrm{X}_3, \qquad v_1 = k_1 (q_1 - x_3)(q_2 - x_3),$$
$$\mathrm{X}_3 \xrightarrow{k_2} \mathrm{X}_1 + \mathrm{X}_2, \qquad v_2 = k_2 x_3 .$$

As far as X_3 is concerned, the stoichiometry matrix S and the reaction rate v

can be written as[2]

$$S = \begin{bmatrix} 1 & -1 \end{bmatrix}, \quad v = \begin{bmatrix} v_1 \\ v_2 \end{bmatrix} = \begin{bmatrix} k_1 (q_1 - x_3)(q_2 - x_3) \\ k_2 x_3 \end{bmatrix}.$$

The concentration $x_3(t)$ of the complex thus evolves according to

$$\frac{dx_3}{dt} = Sv = k_1 (q_1 - x_3)(q_2 - x_3) - k_2 x_3.$$

Example 2.10 (Lotka–Volterra model) Revisit the mutual interactions (2.7) between the prey X_1 and the predator X_2. Let $n_1(t)$ and $n_2(t)$ denote the copy numbers of X_1 and X_2, respectively. The number n_A of the food items A is assumed to be unchanged by consumption during the time scale of interest. The reaction rates according to the mass-action kinetics follow from (2.18) to be (listed to the right)

$$X_1 + A \xrightarrow{\hat{k}_1} 2X_1, \qquad \hat{v}_1 = \hat{k}_1 n_A n_1,$$

$$X_1 + X_2 \xrightarrow{\hat{k}_2} 2X_2, \qquad \hat{v}_2 = \hat{k}_2 n_1 n_2,$$

$$X_2 \xrightarrow{\hat{k}_3} \varnothing, \qquad \hat{v}_3 = \hat{k}_3 n_2.$$

As far as X_1 and X_2 are concerned, the stoichiometry matrix S and the reaction rate v can be written as

$$S = \begin{bmatrix} 1 & -1 & 0 \\ 0 & 1 & -1 \end{bmatrix}, \quad \hat{v} = \begin{bmatrix} \hat{k}_1 n_A n_1 \\ \hat{k}_2 n_1 n_2 \\ \hat{k}_3 n_2 \end{bmatrix}.$$

The ODEs governing the time courses of $n_1(t)$ and $n_2(t)$ can be constructed from the vector Sv as

$$\left.\begin{aligned} \frac{dn_1}{dt} &= \left(\hat{k}_1 n_A - \hat{k}_2 n_2\right) n_1, \\ \frac{dn_2}{dt} &= \left(\hat{k}_2 n_1 - \hat{k}_3\right) n_2. \end{aligned}\right\} \tag{2.23}$$

---

[2]The full stoichiometry matrix for the 3-species 2-reaction scheme has three rows and two columns.

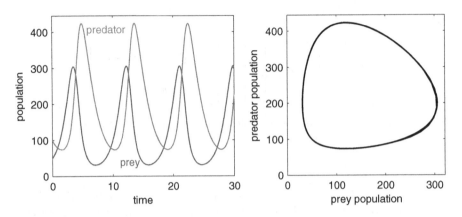

Figure 2.6 Deterministic simulation of the Lotka–Volterra model. *Left*: time course, *Right*: phase plot. Parameters (in sec^{-1}): $\hat{k}_1 = 1$, $\hat{k}_2 = 0.005$, $\hat{k}_3 = 0.6$. Initial population is taken as 50 individuals of prey for 100 individuals of predator.

A numerical solution of the ODEs above is the time plot shown in Figure 2.6 side by side with the associated phase plot.

Example 2.11 (Enzyme kinetic reaction) For the enzyme kinetic reaction (2.8), we write $x_E(t)$, $x_S(t)$, $x_{ES}(t)$, and $x_P(t)$ for the respective time-dependent molar concentrations of E, S, ES, and P. The solution is usually assumed to respect two conservation laws:

$$x_E(t) + x_{ES}(t) = x_E^{tot} \quad \text{and} \quad x_S(t) + x_{ES}(t) + x_P(t) = x_S^{tot}, \qquad (2.24)$$

where x_E^{tot} and x_S^{tot} are, respectively, the total concentrations of the enzyme and substrate determined by the initial conditions. We can choose $x = (x_S, x_{ES})^T$ as the state vector sufficient to describe the system because the remaining two variables can be determined from the conservation relations above. The channelwise mass-action kinetic laws for the reaction scheme (2.8) are (list on the right):

$$
\begin{array}{ll}
E + S \xrightarrow{\ k_1\ } ES, & v_1 = \left(x_E^{tot} - x_{ES}\right) k_1 x_S, \\[1mm]
ES \xrightarrow{\ k_2\ } E + S, & v_2 = k_2 x_{ES}, \\[1mm]
ES \xrightarrow{\ k_3\ } E + P, & v_3 = k_3 x_{ES}.
\end{array}
$$

As far as S and ES are concerned, the stoichiometry matrix S and the reaction

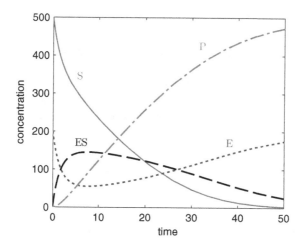

Figure 2.7 Deterministic time course of the enzyme kinetic reaction. Parameters: $k_1 = 10^{-3}\,(\text{nM sec})^{-1}$, $k_2 = 10^{-4}\,\text{sec}^{-1}$, $k_3 = 0.1\,\text{sec}^{-1}$. Initial concentrations: $x_S = 500\,\text{nM}$, $x_E = 200\,\text{nM}$, $x_{ES} = x_P = 0\,\text{nM}$.

rate v can be written as

$$S = \begin{bmatrix} -1 & 1 & 0 \\ 1 & -1 & -1 \end{bmatrix}, \quad v = \begin{bmatrix} (x_E^{\text{tot}} - x_{ES})\,k_1 x_S \\ k_2 x_{ES} \\ k_3 x_{ES} \end{bmatrix}.$$

The concentrations evolve according to the following set of nonlinear coupled ODEs (constructed from the vector Sv)

$$\begin{aligned} \frac{dx_S}{dt} &= k_2 x_{ES} - \left(x_E^{\text{tot}} - x_{ES}\right) k_1 x_S, \\ \frac{dx_{ES}}{dt} &= \left(x_E^{\text{tot}} - x_{ES}\right) k_1 x_S - (k_2 + k_3)\, x_{ES}. \end{aligned} \tag{2.25}$$

A numerical solution of the ODEs above is the time plot shown in Figure 2.7.

Michaelis–Menten kinetics: Following Michaelis and Menten [99] and Briggs and Haldane [19], in addition to the assumption of a constant total enzyme concentration x_E^{tot}, we make an additional assumption that the concentration x_{ES} of the substrate-bound enzyme changes little over time, assuming a quasi steady state, that is,

$$\frac{\mathrm{d}x_{\mathrm{ES}}}{\mathrm{d}t} = \left(x_{\mathrm{E}}^{\mathrm{tot}} - x_{\mathrm{ES}}\right) k_1 x_{\mathrm{S}} - (k_2 + k_3)\, x_{\mathrm{ES}} \approx 0,$$

which is reasonable if the concentration x_{ES} of the substrate-bound enzyme changes much more slowly than those of the product and substrate. The above steady-state assumption can rearranged to form an algebraic expression for the steady-state concentration of the complex:

$$x_{\mathrm{ES}} = \frac{x_{\mathrm{E}}^{\mathrm{tot}} x_{\mathrm{S}}}{\left(\frac{k_2 + k_3}{k_1}\right) + x_{\mathrm{S}}} = \frac{x_{\mathrm{E}}^{\mathrm{tot}} x_{\mathrm{S}}}{K_{\mathrm{M}} + x_{\mathrm{S}}},$$

where $K_{\mathrm{M}} = (k_2 + k_3)/k_1$ is known as the Michaelis–Menten constant. This can be combined with the fact that the product concentration x_{P} changes at the rate

$$\frac{\mathrm{d}x_{\mathrm{P}}}{\mathrm{d}t} = v_3 = k_3 x_{\mathrm{ES}} = \frac{k_3 x_{\mathrm{E}}^{\mathrm{tot}} x_{\mathrm{S}}}{K_{\mathrm{M}} + x_{\mathrm{S}}}.$$

Thus the 3-reaction enzymatic network has been reduced to a single reaction channel S $\rightarrow$ P with reaction rate

$$\frac{\mathrm{d}x_{\mathrm{P}}}{\mathrm{d}t} = -\frac{\mathrm{d}x_{\mathrm{S}}}{\mathrm{d}t} = v\,(x_{\mathrm{S}}) = \frac{v_{\max} x_{\mathrm{S}}}{K_{\mathrm{M}} + x_{\mathrm{S}}},$$

where $v_{\max} = k_3 x_{\mathrm{E}}^{\mathrm{tot}}$ is the initial (maximum) reaction rate.

Example 2.12 (Schlögl model) For the Schlögl reaction scheme (2.9), write x_{A} and x_{B} for the constant respective concentrations of chemicals A and B, and $x(t)$ for the time-dependent concentration of chemical X. The reaction rates according to the mass-action kinetics follow from (2.18) to be (listed on the right)

$$
\begin{aligned}
\mathrm{A} + 2\mathrm{X} &\xrightarrow{k_1} 3\mathrm{X}, & v_1 &= k_1 x_{\mathrm{A}} x^2, \\
3\mathrm{X} &\xrightarrow{k_2} \mathrm{A} + 2\mathrm{X}, & v_2 &= k_2 x^3, \\
\mathrm{B} &\xrightarrow{k_3} \mathrm{X}, & v_3 &= k_3 x_{\mathrm{B}}, \\
\mathrm{X} &\xrightarrow{k_4} \mathrm{B}, & v_4 &= k_4 x.
\end{aligned}
$$

As far as X_3 is concerned, the stoichiometry matrix S and the reaction rate v

can be written as

$$S = \begin{bmatrix} 1 & -1 & 1 & -1 \end{bmatrix}, \quad v = \begin{bmatrix} k_1 x_A x^2 \\ k_2 x^3 \\ k_3 x_B \\ k_4 x \end{bmatrix}.$$

The deterministic ODE turns out to be

$$\frac{dx}{dt} = Sv = k_1 x_A x^2 - k_2 x^3 + k_3 x_B - k_4 x. \tag{2.26}$$

Example 2.13 (Stochastic focusing) The branched reaction scheme (2.10):

$$\varnothing \underset{k_d}{\overset{k_s}{\rightleftharpoons}} S, \quad \varnothing \underset{k_a x_S}{\overset{k_i}{\rightleftharpoons}} I \overset{k_p}{\longrightarrow} P \overset{1}{\longrightarrow} \varnothing.$$

Write $x_S(t)$, $x_I(t)$, and $x_P(t)$ for the respective time-dependent molar concentrations of the signal S, the intermediary precursor I, and product P. The reaction rates based on mass-action kinetics are k_s for synthesis of S and $k_d x_S$ for its degradation, k_i for synthesis of I and $k_a x_S x_I$ for its degradation, $k_p x_I$ for the I $\rightarrow$ P conversion and $-x_P$ the product degradation. Ordering the species as {S, I, P}, the stoichiometry matrix S and the reaction rate v take the forms

$$S = \begin{bmatrix} 1 & -1 & 0 & 0 & 0 & 0 \\ 0 & 0 & 1 & -1 & -1 & 0 \\ 0 & 0 & 0 & 0 & 1 & -1 \end{bmatrix}, \quad v = \begin{bmatrix} k_s \\ k_d x_S \\ k_i \\ k_a x_S x_I \\ k_p x_I \\ -x_P \end{bmatrix}.$$

The deterministic system of ODEs for the system can now be read from the vector Sv:

$$\left.\begin{array}{l} \dfrac{dx_S}{dt} = k_s - k_d x_S, \\[2mm] \dfrac{dx_I}{dt} = k_i - (k_p + k_a x_S)\, x_I, \\[2mm] \dfrac{dx_P}{dt} = k_p x_I - x_P\,. \end{array}\right\} \qquad (2.27)$$

Example 2.14 (Hyperbolic control) If the pool of I-molecules is insignificant, the two reactions involving their loss are fast enough, and X_S does not change significantly during the life span of an individual I-molecule, then we can assume the steady state of ending up in P or A to be reached immediately. The steady-state abundance of I-molecules, obtained by setting to zero the right side of the second equation in (2.27), is $x_I^{ss} = k_i/(k_p + k_a X_S)$. That leads to the following simplification of (2.27):

$$\left.\begin{array}{l} \dfrac{dx_S}{dt} = k_s - k_d x_S, \\[2mm] \dfrac{dx_P}{dt} = \dfrac{k_p k_i}{k_p + k_a x_S} - x_P, \end{array}\right\} \qquad (2.28)$$

and a corresponding reduction of the branched reaction scheme (2.10):

$$\varnothing \underset{k_d}{\overset{k_s}{\rightleftharpoons}} S, \qquad \varnothing \underset{1}{\overset{k_i/(1 + x_S/K)}{\rightleftharpoons}} P, \qquad (2.29)$$

where $K = k_p/k_a$ is the inhibition constant. The denominator $1 + x_S/K$ in the expression for the new effective rate coefficient suggests the name "hyperbolic control" for the product molecule by the signal molecule.

Example 2.15 (Gene regulation) For the gene regulation scheme (2.11):

$$G \xrightarrow{k_m} G + M \qquad \text{(transcription)},$$

$$M \xrightarrow{k_p} M + P \qquad \text{(translation)},$$

$$G + P \underset{k_u}{\overset{k_b}{\rightleftharpoons}} GP \quad \text{(binding/unbinding)},$$

$$M \xrightarrow{k_m^-} \varnothing, \; P \xrightarrow{k_p^-} \varnothing \qquad \text{(degradation)},$$

write $x_M(t)$, $x_G(t)$, and $x_P(t)$ for the respective time-dependent molar concentrations of mRNA M, the unbound gene G, and protein P. The total gene concentration x_G^{tot} is assumed to be constant, so that the bound (repressed) protein concentration is simply $x_G^{tot} - x_G$. The reaction rates based on mass-action kinetics are $k_m x_G$ for transcription, $k_p x_M$ for translation, $k_b x_G x_P$ for the gene–protein binding, $k_u \left(x_G^{\text{tot}} - x_G\right)$ for the gene–protein unbinding, $k_m^- x_M$ for mRNA degradation, and $k_p^- x_P$ for protein degradation. Ordering the species as $\{M, G, P\}$, the stoichiometry matrix S and the reaction rate v take the forms

$$
S = \begin{bmatrix} 1 & 0 & 0 & 0 & -1 & 0 \\ 0 & 0 & -1 & 1 & 0 & 0 \\ 0 & 1 & -1 & 1 & 0 & -1 \end{bmatrix}, \quad v = \begin{bmatrix} k_m x_G \\ k_p x_M \\ k_b x_G x_P \\ k_u \left(x_G^{\text{tot}} - x_G\right) \\ k_m^- x_M \\ k_p^- x_P \end{bmatrix}.
$$

The deterministic system of ODEs for the system can now be constructed from the vector Sv:

$$
\left.\begin{aligned}
\frac{\mathrm{d}x_M}{\mathrm{d}t} &= k_m x_G - k_m^- x_M, \\
\frac{\mathrm{d}x_G}{\mathrm{d}t} &= k_u \left(x_G^{\text{tot}} - x_G\right) - k_b x_G x_P, \\
\frac{\mathrm{d}x_P}{\mathrm{d}t} &= k_p x_M + k_u \left(x_G^{\text{tot}} - x_G\right) - \left(k_b x_G + k_p^-\right) x_P.
\end{aligned}\right\} \tag{2.30}
$$

2.4 The Art of Modeling

To do mathematical modeling at the life sciences interface is to engage in an act of discovery and conjecture. The art of modeling is not in the accuracy of a mathematical model but in the explanation, that is, in the argument that is developed in the process outlined in Figure 1.4. It is this argument and its context that give the model its validity. Mathematical modeling of cell-biological systems is an art—the art of asking suitable questions, choosing an appropriate conceptual framework to formulate and test hypotheses, and making appropriate assumptions and simplifications. Our goal is to improve the understanding of living systems, and we believe that there is nothing more practical in addressing the complexity of living systems than mathematical modeling.

What we are seeking is an understanding of the functioning of cells, of their behavior and the mechanisms underlying it. When we speak of mechanisms and principles as being the goal of our scientific quest, we really mean that we are interested in the system's organization [168]. In living systems there are two forms of interlinked organization: The *structural organization* of a cell refers to the arrangement and structural (material or biophysical) properties of its components— organelles and macromolecules. Inseparable from the cell's structural organization is its *functional organization*, describing the processes that determine the cell's behavior or '(mal)functioning'. Interacting with other cells and/or its environment, the cell realizes four key functions: growth, proliferation, apoptosis, and differentiation. The processes that realize these functions of a cell can be further organized into three process levels: gene regulation, signal transduction, and metabolism (Figure 1.3). The experimental study of any one of these cell functions and any one of these process levels is subject to high degrees of specialization. These specialized research fields are often separated by technology, methodology, and culture. This depth of specialization is a hurdle to a comprehensive understanding of how cells and cell populations (mal)function.

In summary, systems theory is the study of organization, using mathematical modeling. With respect to systems biology, the key challenges are:

- Depending on the data and question at hand, what approach to choose and why?

- How do I decompose a complex system intro tractable subsystems?

- Given an understanding of subsystems, how can one integrate these data and models into an understanding of the system as a whole?

Techniques for coupling/embedding models of components built on disparate time and length scales, and often with different modeling techniques, into larger models spanning much longer scales are in their infancy and require further investigation. We limit ourselves in this text to a small subset of these challenges and focus on one particular approach to studying small subsystems.

Problems

2.1. When the volume is not known or important, it is convenient to choose a value so that each nanomolar concentration is numerically equal to the corresponding copy number. Compute that value of the volume.

2.2. Suppose species concentration is measured in molecules per μm^3 (cubic micrometers) of volume. What can you say about the magnitude and unit of the system size?

2.3. Consider the irreversible bimolecular reaction

$$A + B \xrightarrow{\;\;k\;\;} X + Y.$$

Temporal changes in species concentration for this reaction are restricted by a conservation relation.

1. Write down the conservation relation for concentrations in terms of initial concentrations.

2. Express the reaction rate law in terms of time-dependent concentration of X.

3. Implement the rate law as a Matlab function handle. Assume that $k = 1\,\mathrm{sec}^{-1}$ and initial abundances are $2\,\mathrm{M}$ for A, $3\,\mathrm{M}$ for B, and $0.5\,\mathrm{M}$ for X.

4. Call the function handle in an ODE solver to compute and plot the time-course concentration of X for the first 5 seconds.

2.4. Consider the consecutive reaction

$$X_1 \xrightarrow{\;\;k_1\;\;} X_2 \xrightarrow{\;\;k_2\;\;} X_3.$$

1. Write down the differential equation for the concentration X_2.

2. Assume zero initial concentrations except for the first reactant, which is $10\,\mathrm{M}$, and take $1\,\mathrm{sec}^{-1}$ for both rate constants. Run the following script:

```
x0 = [10;0]; % initial concentrations
k1 = 1; k2 = k1; % rate constants
v = @(t,x) [-k1*x(1);k1*x(1)-k2*x(2)]; % rate law
[t,x] = ode45(v,[0 5],x0); % solver
plot(t,x(:,2)) % plot x2
```

Repeat the simulation for $k_2 = 0.1k_1$ and $k_2 = 10k_1$. Relate the relative magnitudes of the rate constants to the relative reaction rates.

3. If one of the two reactions is much faster than the other, the overall reaction rate is determined by the slower reaction, which is then called the "rate-determining step." For each value of k_2, which reaction is rate-determining?

2.5. Recall the rate law

$$v(x) = k \prod_{i=1}^{s} x_i^{g_i}$$

with definitive orders for a chemical reaction. It can be implemented as a function handle:

```
k = 2; % rate constant
g = [0 1 1 0 0 1]'; % reaction stoichiometry
v = @(x) k*prod(x.^g); % rate law
```

for the specified values of k and g.

1. Evaluate the rate expression for

$$x = \begin{bmatrix} 2 & 0.5 & 0 & 1.5 & 0 & 3 \end{bmatrix}^T.$$

What problem did you encounter? Can you figure out why?

2. Reimplement the rate law as a function that accounts for the pitfall you encountered.

2.6. Consider a simple network

$$2X_1 \xrightarrow{k_1} X_2, \quad X_2 + X_3 \xrightarrow{k_2} X_4,$$

of metabolites. The metabolite concentrations are measured in molecules per μm^3 (cubic micrometers).

1. Set up the stoichiometry matrix S.

2. Write down the expression, based on mass-action kinetics, for the two reaction rates v_1 and v_2 in terms of species concentrations.

3. How would you combine the two results to construct the ODEs that describe how species concentrations change with time.

4. Complete the following script based on the quantities in the above steps in order to compute and plot the species concentrations against time over 500 seconds:

```
% initial abundance (molecules per cubic micrometer)
x0 = [10;0;5;0];
% rate constants (per cubic micrometer per second)
k = [1e-3;3e-3];
```

```
% S = ?; % stoichiometry matrix
% v = @(x) ?; % rate law
% dxdt = ?; % ODEs
[t,x] = ode15s(v,[0 500],x0); % solver
plot(t,x) % plot x
```

5. Discover the conservation relations in the reaction scheme and utilize them to rewrite the rate equations so that they involve concentrations of X_2 and X_3 only.

6. Modify the code accordingly and check the result by plotting and comparing with the previous implementation.

2.7. The repressilator consists of three genes connected in a feedback loop such that each gene product represses the next gene in the loop and is repressed by the previous gene [40]. If we use subscripts $i = 1$, 2, 3 to denote the three genes; M_i represents mRNAs, and P_i the proteins. The gene network can be represented by the reaction scheme

$$M_i \underset{\alpha(M_{i-1})}{\overset{1}{\rightleftharpoons}} \varnothing, \quad P_i \xrightarrow{b} \varnothing, \quad M_i \xrightarrow{b} M_i + P_i$$

where i runs through 1, 2, 3 and $P_0 = P_3$. For simplicity, assume relative (nondimensional) concentrations. The mRNA transcription rate is

$$\alpha(x) = a_0 + \frac{a_1}{(1 + x)^h},$$

where a_0 is the transcription rate in the presence of saturating repressor and $a_0 + a_1$ represents the maximal transcription rate in the absence of the repressor. The exponent h in the denominator is the Hill coefficient. The parameter b appears as the protein degradation rate constant and translation rate constant.

1. Set up the stoichiometry matrix S by adopting the ordering M_1, M_2, M_3, P_1, P_2, P_3 for species and the ordering $M_1 \rightarrow \varnothing$, $M_2 \rightarrow \varnothing$, $M_3 \rightarrow \varnothing$, $P_1 \rightarrow \varnothing$, $P_2 \rightarrow \varnothing$, $P_3 \rightarrow \varnothing$, $\varnothing \rightarrow M_1$, $\varnothing \rightarrow M_2$, $\varnothing \rightarrow M_3$, $M_1 \rightarrow M_1 + P_1$, $M_2 \rightarrow M_2 + P_2$, $M_3 \rightarrow M_3 + P_3$ for reactions.

2. Write down the expressions for channelwise reaction rates v_j in terms of species concentrations.

3. Combine the two results to construct the ODEs that describe how species concentrations change with time.

4. Complete the following script based on the quantities in the above steps in order to compute and plot the protein levels for 50 time units:

```
% parameters
a0 = 0.25; a1 = 250; b = 5; h = 2.1;
% S = ?; % stoichiometry matrix
% v = @(x) ?; % rate law
dxdt = @(t,x) S*v(x); % ODEs
tmax = 50; % time
x0 = [0 0 0 4 0 15]'; % initial concentration
[t,x] = ode45(dxdt,[0 tmax],x0); % solution
plot(t,x(:,4:6)) % plot protein levels
```

5. Do you see oscillations in the protein levels? Play with the parameter values and initial conditions to see whether you always get oscillations.

6. Looking at time plots for checking oscillations is one way to solve part 5 above. An alternative is to look at the phase plot. Extend the code to plot the phase plots for each mRNA–protein pair. What do these phase plots reflect?

2.8. The repressilator model in the last exercise is a nondimensional version of the original model available on the biomodel database http://biomodels. caltech.edu/BIOMD0000000012. Run the online simulation provided. Do you see oscillations in the protein levels? Play with the parameter values and initial conditions to see whether you always get oscillations.

Chapter 3

Randomness

This chapter is intended to provide an informal introduction to concepts that are necessary for stochastic modeling. The goal is to prepare readers unfamiliar with probability theory for the next chapter, which will be a more formal discourse. It is recommended to read this chapter before reading the next one.

3.1 Terminology

The terms "noise," "fluctuations," and "randomness" can be confusing because they are not clearly defined for all systems. Loosely speaking, fluctuations or noise refers to the irregular aspects of observations made of a phenomenon representing various disturbances over which there is no clear control. It also refers to the factors (sources) responsible for the fluctuations. It has often negative associations as something undesirable, something that should be removed or avoided. In biology, however, noise can also contain information about details in a system under study and thus plays an important role. In this text the term (noise) is used with the understanding that it may well be something desirable. The term "randomness" refers to the fact that we have no knowledge about, or have no control over, the factors responsible for the irregularities and lack of pattern in the observations. The term has been explained beautifully in [29] which we will adopt. What does it mean to say that the time of occurrence of an event, or between two events, is random? Examples are the time of a specific gene mutation or between completion of transcription and initiation of translation. The first thing to know is that randomness is related not to the phenomena but rather in how we treat phenomena in our investigation. Every scientific investigation looks for relations between objects referred to as measurements. A single measurement takes the form of a real (or a logical) number. However, we are interested in measurements of recurring phenomena, and the objects to be related take the form of abstract symbols, called "variables." These represent measurements, such as time and protein abundance. Let us fix these ideas with the help of our first example, the standard modification, introduced in the previous chapter.

3.1.1 *Representations of Measurements*

Recall the reversible transformations in the reaction scheme (2.5) of a protein between two forms. Consider the time τ between successive modifications (i.e., reaction events). For a specific instance, τ represents a single measurement and takes the form of a real number. However, our interest is not with a single observation of time but rather with the class of measurements. Thus, the time between successive modifications varies depending on many conditions within the cell, and the time is represented as a random variable, which we will denote by T. Here, the single (unique) value τ has been replaced by a range of values, and the adjective "random" renders to the term "variable" a new mathematical meaning, which can be thought of as a rule (or mapping) that assigns a unique real number to each measurement of the quantity of interest, in this case time between two transformations. This (rule of) assignment cannot be arbitrary but must satisfy some requirement in an axiomatic framework. In that yet to be introduced framework, it is said that T is a "measurable function" from a "probability space" into the space of real numbers. Our purpose in this chapter will be to establish and clarify these spaces. Since real numbers are assigned to measurements by the random variable T, it is said to be a *continuous* random variable, in contrast to a *discrete* random variable, which assigns values from a finite, or countably infinite, collection. For instance, the copy number $n(t)$ of unmodified protein molecules, at time t, is an integer in a specific measurement. If we track this abundance for a single cell, we get a time function that is deterministic, the latter meaning, by definition, that there is a certain value at each time point. However, we are typically interested in the behavior of the proteins for an arbitrary cell, and then, the measurement is not deterministic but varies from cell to cell. In this case, the measurement is represented as a time-dependent random variable, denoted by $N(t)$, again the word "random" appearing as part of the mathematical term "random variable." The deterministic variable $n(t)$ takes integer values. The random variable $N(t)$ is a function (or a rule) that assigns a unique integer, from 1 to n^{tot}, to each measurement of the abundance. In the axiomatic terminology to be introduced latter, $N(t)$ is a measurable function from a probability space to the numerical space of integers 1 to n^{tot}. Whereas $n(t)$ is referred to as a "time function," $N(t)$ is referred to as a "random time function," a "random time process," or a "stochastic process". In every instance in which the word "random" is used, it requires a definition in terms of the underlying mathematical spaces. None of this makes any suppositions concerning things-in-themselves. In the context of science, "random" is simply a word adopted by mathematics and defined therein within the framework of axiomatic probability theory.

3.1.2 Sources of Noise

The classification of noise (e.g., internal/external) can have different meanings for different systems. However, the various kinds of noise in gene expression have been clearly defined in [13, 30, 74, 117, 128]. Intrinsic noise has origins within the boundaries of the system under consideration and arises from the very discrete nature of the chemical events of gene expression. Extrinsic noise arises from interactions of the (sub)system with other (sub)systems or the outside environment. Following [128], noise in gene expression refers to the stochastic variation of an (expressed) protein concentration within isogenic cells having the same history and conditions (environment). Placing two gene reporters in the same cell and quantifying their gene expression (by the abundance of their target proteins) allows the following categorization of noise (see Figure 2 in [128]). *Intrinsic noise* arises from sources that create differences (in the gene expression) between the two reporters in the same cell, and *extrinsic noise* arises from sources that have equal effects on the two reporters in the same cell but create differences between two cells. Stochastic events during gene expression will then emerge as intrinsic noise, whereas differences between cells will appear as extrinsic noise. Extrinsic noise can be *global* when fluctuations in basic reaction rates affect expressions of all genes, or it can be *pathway-specific*. It is important to realize that extrinsic noise can be theoretically isolated from the system, but intrinsic noise is the very essence (discrete nature) of the underlying molecular events and cannot be separated (even hypothetically) from the system.

3.2 Probability

Again, we use the reversible transformations of a protein between two forms in the reaction scheme (2.5). Imagine that every protein molecule, irrespective of its form, has been tagged with a unique label. Representing the labels by integers 1 to n^{tot}, we can represent the whole collection of molecules with the set

$$\Omega = \left\{1, 2, \ldots, n^{tot}\right\} .$$

Your first distribution: Now imagine a thought experiment of picking at random (i.e., without considering its form or spatial location) one protein molecule from the lot. We can think of it as a *random experiment* because we cannot uniquely determine the outcome. The *outcome* of this experiment will be an element i of the above set Ω. We can think of each element i as a *sample point* in a *sample space* Ω in the same sense that a number is considered to be a point in some numerical space (e.g., real line, complex plane, vector space). Logically, every molecule is as likely to be picked as any other molecule. So if we assign a number $0 \leq p_i \leq 1$ to quantify the chance or likelihood of the molecule labeled i being picked, then $p_i = 1/n^{\mathrm{tot}}$ for all ω. If you have ever heard the term "probability" before, it is natural for you to anticipate that the number p_i should be a probability. Well, you have to wait a bit.

3.2.1 Probability of What?

When one first encounters probabilities, it is tempting to think that a probability is assigned, like the number p_i, to the sample points i. This may seem logical at first, but it suffers from a caveat. Since probability is meant to be a measure of chance, we should be able to assign probabilities, in a consistent manner, to a variety of questions (of interest) about the outcome. We cannot limit such questions to the sample points (individual elements) only. For instance, we could be interested in the probability of picking a molecule with a label $i < n^{\mathrm{tot}}/2$ that corresponds to a subset of the sample space Ω. It turns out that any question we could ask about the outcome can be translated to whether the outcome is an element of some subset A of the sample space Ω. Such a subset can be thought of as an event. This perfectly matches the use of the term "events" in everyday life. We call something an event only if it is interesting to us. Therefore, it is more logical to assign probabilities to an event, which is a subset A of the sample space. The probability that has been assigned to an event A is usually written as $\Pr[A]$. The sample points, the individual elements i, can also be represented as single-element events of the form $\{i\}$, which can be thought of as *elementary events*. Now we can say that the equal numbers $p_i = 1/n^{\mathrm{tot}}$ are assigned to the elementary events $\{i\}$, and not to the elements i of the sample space. In other words, $\Pr[\{i\}] = 1/n^{\mathrm{tot}}$ with a caution that $\Pr[i]$ has no meaning.

3.2.2 Event Space

A common misunderstanding about events needs to be removed at this point. We saw that every event can be cast as a subset of the sample space. Does that mean that every subset of the sample space is an event? The answer is, in general, no. Remember that we want to assign probabilities in a consistent framework and would like to guarantee that the desired properties such as the additive rule, and others not discussed so far, always hold. This limits the choice of subsets, because it may not always be possible to assign probabilities with all the desired properties. One obvious example arises when the sample space is infinite: you can't assign a nonzero probability to each elementary outcome, because the sum of all has to be unity. Furthermore, we may not be interested in every subset of Ω, nor do we have to. All this means that, similar to a sample space, we need an *event space* $\mathcal{A}$ that contains all our events of interest. So, while an outcome is an element ω of the sample space

Ω, an event A is an element of an event space $\mathcal{A}$. You can think of an event space as a reduced collection that is short-listed from all possible subsets of the sample space. This also brings another issue to the forefront. The rule of assigning probabilities (real numbers from 0 to 1) to events from an event space $\mathcal{A}$ can be interpreted as a set function $\Pr : \mathcal{A} \to [0, 1]$ that maps the event space to the unit interval on the real line.

3.2.3 This OR That

Many questions asked together can be cast as a new complex question and vice versa. Therefore, events can be combined logically to form new events. For instance, the event $A = \{1, 2\}$ can be written as a set union: $\{1\} \cup \{2\}$. Common sense will guide you that the probabilities should add:

$$\Pr[A] = p_1 + p_2 = 2/n^{\text{tot}},$$

because of greater number of possibilities (sample points). Another example is the outcome $B = \{2, 3\}$ with probability

$$\Pr[B] = p_2 + p_3 = 2/n^{\text{tot}}.$$

This additive rule of probabilities cannot always be used. Now consider the outcome $A \cup B = \{1, 2, 3\}$. We cannot simply add the probabilities as before, since A and B have the sample point 2 in common, and hence p_2 will be added twice. Whenever the set intersection $A \cap B$ is nonempty, there will be such common elements, and the additive rule of probabilities will not hold. Therefore, probabilities can be added only if the individual events have nothing in common. Formally speaking, they are *mutually exclusive* or *disjoint*: $A \cap B = \emptyset$. When that is not the case, a little thought about the matter could convince you that the probability $\Pr[A \cap B]$ should be subtracted from the sum of probabilities to give the correct probability.

3.3 Random Variables

So far, in this standard-modification example, our discussion about protein molecules disregarded the form. Now imagine that in addition to the previous labeling, groups of molecules in the two forms are tagged with two different labels, which we represent here by 1 for unmodified and 0 for modified. This new labeling will not change anything regarding the assignment of probabilities to the outcomes of our previous experiment as long as we do not ask a question that relates to the type or form of the molecule. If we do ask about which group a molecule belongs to, the answer can be found only after selecting the molecule and observing whether it has been labeled 1 or 0. In other words, we need to represent the form (or state) of a molecule by a random variable Y that assigns the value (1)0 to the observation if the protein molecule is (un)modified. It is important to realize that Y assigns a value to a whole group of molecules. Thus Y maps all the labels from the previous tagging, which have been newly labeled (1)0, to a single point (1)0 on the real line. Now you can guess why the random variable is considered a function that maps the sample space to points on the real line.

Suppose we know, before picking a protein molecule, that n molecules of the lot are unmodified. This means that $n^{tot} - n$ molecules are modified. This knowledge will change the probability of any event (question) that is asked about the type of molecules. Let us consider the event A that a selected molecule is labeled 1. The event A can also be written, in a more intuitive notation, as $Y = 1$, with the understanding that it does not represent the value 1 of Y, but a subset A in the event space. Since n molecules, and hence labels, from the previous tagging constitute the event $Y = 1$, its probability is, following the additive rule, the sum of n probabilities, all equal to $1/n^{tot}$, that is,

$$\Pr\left[Y = 1\right]_n = \frac{n}{n^{tot}}, \tag{3.1}$$

the fraction of molecules that are unmodified. The subscript n here reflects the condition that n molecules of the lot are unmodified. It is not the same as the probability $\Pr\left[Y = 1\right]$, which is unconditional. At this point we do not know the expression for the latter. We will refer to the yet to be determined expressions for $\Pr\left[Y = y\right]$ as p_u in the discussion to follow. In other words, p_u is the probability for a molecule to be an unmodified protein and $p_w = 1 - p_u$ is the complementary probability of being a modified protein.

Relative size: The probability (3.1) is a ratio of numbers $\#A = n$ and $\#\Omega = n^{tot}$ of elements in the event space and the sample space. Thus, the probability of an event A can be regarded as a measure of the size of A relative to Ω.

Your first distribution: The description of the random variable Y is complete, because we know all the possible values, 0 and 1, and the probabilities that those values will be assigned in a measurement (the selection of a molecule here). Such a description is referred to as a *probability distribution*, a collection of all possible values, assigned by the random variable to individual measurements, together with the associated probabilities. The particular probability distribution of Y here can be expressed as

$$\Pr\left[Y = 1\right] = p_u, \quad \Pr\left[Y = 0\right] = 1 - p_u$$

where y is a possible value of the measurement. This is so-called *Bernoulli distribution*, and Y is said to be a *Bernoulli variable*, with *success probability* p_u. In this terminology, $Y = 1$ is a *success* and $Y = 0$ a *failure*. The detailed description of a distribution is either not always possible to know, or may be difficult to interpret due to overwhelming complexity. Therefore, summarizing features including the average (or mean) and variance are used to characterize the distribution.

3.3.1 Average

The average is a measure of the expected value of a distribution. In everyday life, the average is usually associated with a list of values of some quantity and is computed as the sum of all values in the list divided by the number of values. The sum in this computation is usually a weighted sum because some values may be repeated. For our current example, the random variable Y assigns the value $y = 1$ to all unmodified molecules. Thus, if n molecules of the lot are known to be unmodified, the weighted sum has the value $y = 1$ added n times and the value $y = 0$ added $n^{tot} - n$ times. The conditional average value of Y then has the expression

$$\langle Y \rangle_n = \frac{(1)n + (0)(n^{tot} - n)}{n^{tot}} = \frac{n}{n^{tot}},$$

which, from (3.1), is equal to the conditional probability of success. It follows that the unconditional average is equal to the (unconditional) success probability p_u. An interesting pattern appears if we write the unconditional version of the above equation:

$$\langle Y \rangle = (1)p_u + (0)p_w = p_u,$$

which is a weighted sum in which each value y is weighted by the probability $\Pr[Y = y]$. This procedure can be extended to find an expression for the average value of any discrete random variable Y. The resulting expression for the average is

$$\langle Y \rangle = \sum_y y \Pr[Y = y] . \tag{3.2}$$

The averaging operation expressed here is mathematically known as the *expectation*.

3.3.2 Variance

The variance is a measure of the spread of the distribution around the average. The variance is the average of $(Y - \langle Y \rangle)^2$, the random variable representing the squared deviation from the average value. Replacing, in the above expression for the average, Y by $(Y - \langle Y \rangle)^2$ followed by the associated replacement for the sample values y, we get the expression for the variance

$$\langle \delta Y^2 \rangle = \left\langle (Y - \langle Y \rangle)^2 \right\rangle = \sum_y (y - \langle Y \rangle)^2 \Pr[Y = y] . \tag{3.3}$$

The squared deviation is used because the simple deviation $Y - \langle Y \rangle$ has a zero average, and the use of absolute deviation is mathematically unpleasant. For this example, Y has variance

$$\langle \delta Y^2 \rangle = (1 - p_u)^2 \, p_u + (0 - p_u)^2 \, (1 - p_u) = (1 - p_u) \, p_u .$$

Let us highlight these two important results in words:

> *A Bernoulli random variable with success probability p_u has average value p_u and variance $(1 - p_u) \, p_u$.*

3.3.3 Continuous Random Variable

The relative size interpretation of probability is very useful in cases in which probabilities cannot be assigned to elementary events. For instance, consider a protein that can change its length l between two fixed real numbers a and b. Imagine an experiment in which we measure the protein length L formulated as a random variable that attains (real) values between a and b. The sample space Ω for this experiment is the segment between a and b of the real line. Since the number of sample points, possible length values, is infinite, we cannot assign nonzero probabilities to events of the form $L = l$ because of the requirements of the additive rule. However, we can employ the relative size interpretation of probability. Using the length $b - a$ of the interval as a measure of size of the sample space, a probability can be assigned to any event that takes the form of a subinterval $l \leq L \leq l + \Delta l$. If all subintervals of equal length are equiprobable, then the probability of such a subinterval is its size relative to $b - a$, that is,

$$\Pr[l \leq L \leq l + \Delta l] = \frac{\Delta l}{b - a} .$$

If we shrink the interval length Δl to zero, we can think of $1/(b-a)$ as a *probability density*. In this particular case, the density is constant, and such a distribution is said to be *uniform*. In general, the *probability density function* $p(x)$ of a continuous random variable X is defined in such a way that the probability that X would fall in an interval $[a, b]$ is the integral

$$\Pr[a < X < b] = \int_a^b p(x)\mathrm{d}x.$$

Since a summation in the discrete setting becomes an integral in the continuous setting, the average value of X can be written, in analogy to (3.2),

$$\langle X \rangle = \int_x xp(x)\mathrm{d}x,$$

where the integration is done over all sample values x. Similarly, the variance is a continuous analogue of (3.3):

$$\langle \delta X^2 \rangle = \int_x (x - \langle X \rangle)^2 p(x)\mathrm{d}x.$$

3.3.4 This AND That

Another possible confusion arises in dealing with joint events, which are intersections of the form $A \cap B$. If the two events A and B have nothing in common, they are disjoint, $A \cap B = \emptyset$, that is, an outcome of the experiment cannot be an element of both. In other words, the occurrence of one of the two events rules out the possibility of the other. For instance, in the standard modification example, the two events $Y = 1$ and $Y = 0$ are disjoint because only one of the two can occur in a single selection. However, extending the experiment to selecting two molecules changes the situation. Represent the form of the first selected molecule by Y_1 and that of the second by Y_2. Since we are now dealing with pairs of molecules, the new sample space is the set $\Omega^2 = \Omega \times \Omega$ with all possible pairs of labels. While the events for the same pick are disjoint, $Y_i = 0 \cap Y_i = 1 = \emptyset$, those for alternative picks are not. The joint event $Y_1 = y_1 \cap Y_2 = y_2$ for the alternative picks has size

$$\#Y_1 = y_1 \cap Y_2 = y_2 = \#(Y_1 = y_1)\#(Y_2 = y_2)$$

if the first selected molecule is placed back in the lot before the second selection. This assumption, which makes sense because you would not throw a molecule out after observing its form, makes sure that the sizes of the events and the sample space do not change from the first selection to the second. Following the relative size interpretation of probability, we divide the above sizes by the size $\#\Omega^2 = (\#\Omega)^2$ of the joint sample space, and get

$$\Pr\left[Y_1 = y_1 \cap Y_2 = y_2\right] = \frac{\#\left(Y_1 = y_1\right)\#\left(Y_2 = y_2\right)}{\#\Omega\#\Omega}$$
$$= \Pr\left[Y_1 = y_1\right]\Pr\left[Y_2 = y_2\right].$$

Based on this expression, the selection sequences 11, 00, 10, and 01 have respective probabilities p_u^2, $(1 - p_u)^2$, $p_u\left(1 - p_u\right)$, and $(1 - p_u)\,p_u$; where p_u is the success probability (fraction of molecules that are unmodified). That the probability of a joint event can be written as the product of probabilities of the joined events is exactly what is meant by *independence* of events. Two events A and B are independent if the probability of their joint occurrence is the product of their individual probabilities. Note that in this example this independence is a consequence of our assumption that the first molecule, after being selected, is placed back in the lot before the second selection. When this is not true, for example if the two molecules are selected together, the sample size is reduced for the second selection and the joint sample space has the reduced size $n^{\text{tot}}\,(n^{\text{tot}} - 1)$. Moreover, the sizes of the joint events depend on the outcome of the first selection. Without getting into details of how the probabilities should be assigned in this case, we can conclude that in general, the probability of the joint occurrence of events A and B is

$$\Pr\left[A \cap B\right] = \Pr\left[B\right]\Pr\left[A \,|\, B\right],$$

where the new notation $\Pr\left[A \,|\, B\right]$ denotes the probability of the event A after it is known that the event B has occurred. It is a conditional probability in which the occurrence of A is conditioned on B. When the two events are independent, knowledge about the occurrence of B has nothing to do with the occurrence of the event A. In that case, $\Pr\left[A \,|\, B\right] = \Pr\left[A\right]$, and the joint probability can be factored: $\Pr\left[A \cap B\right] = \Pr\left[A\right]\Pr\left[B\right]$.

3.3.5 Total Law of Probability

The factorization of the joint probability $\Pr\left[A \cap B\right]$ has useful consequences. If the sample space Ω can be partitioned into a disjoint collection B_1, B_2, ..., then $A \cap B_i$ represents the joint occurrence of A and B_i. Summing up probabilities of these joint occurrences naturally gives the probability of A. Mathematically, this assumes the form

$$\Pr\left[A\right] = \sum_i \Pr\left[A \cap B_i\right] = \sum_i \Pr\left[B_i\right]\Pr\left[A \,|\, B_i\right],$$

known as the "total law of probability." It essentially says that the probability
of an event can be represented as a sum of probabilities of its joint occurrences
with a collection of disjoint events that span the whole sample space.

Success probability: The total law of probability allows us to find the
expression for the success probability p_u. Recall that the success probability
(3.1) is conditional on the knowledge that n molecules of the lot are unmodified.
If we represent the copy number of unmodified proteins by the random variable
N, then the aforementioned condition can be represented by the event $N = n$.
The event $Y = 1$ can be thought of as a union of joint events $Y = 1 \cap N = n$
for all values n. The law of total probability for this case is

$$\Pr\left[Y = 1\right] = \sum_n \Pr\left[Y = 1 \cap N = n\right] = \sum_n \Pr\left[N = n\right]\Pr\left[Y = 1\right]_n .$$

Using the expression (3.1) for the conditional success probability in the above
will turn it into

$$p_u = \Pr\left[Y = 1\right] = \frac{\langle N \rangle}{n^{\text{tot}}}, \tag{3.4}$$

where the average copy number $\langle N \rangle$ is recognized following (3.2). Thus the
success probability is the average fraction of molecules in the unmodified form.
In other words, the fraction $n(t)/n^{\text{tot}}$ has been replaced by the average fraction
when the condition $N = n$ is removed from the success probability (3.1).

3.3.6 Sum of Random Variables

The sum of two (or more) random variables is also a random variable. In
relation to the above discussion, the sum $Y = Y_1 + Y_2$ is a random variable that
represents the number of successes. Thus $Y = 0$ occurs with the (selection)
sequence 00, $Y = 1$ occurs with the sequences 10 and 01, and $Y = 2$ occurs
with the sequence 11. Since we know the probabilities of those sequences, the
probability distribution of the random sum Y can be expressed as

$$\Pr\left[Y = y\right] = \begin{cases} (1 - p_u)^2 & \text{if } y = 0, \\ 2p_u\left(1 - p_u\right) & \text{if } y = 1, \\ p_u^2 & \text{if } y = 2, \end{cases}$$

where the second case $y = 2$ requires the summation of probabilities of the
corresponding sequences 10 and 01. The random sum has average value

$$\langle Y \rangle = (0)\left(1 - p_u\right)^2 + (1)2p_u\left(1 - p_u\right) + 2p_u^2 = 2p_u,$$

and variance

$$\langle \delta Y^2 \rangle = (0 - 2p_u)^2 (1 - p_u)^2 + (1 - 2p_u)^2 2p_u (1 - p_u) + (2 - 2p_u)^2 p_u^2$$
$$= 2p_u (1 - p_u) \ .$$

The point to note here is that the average of the random sum is the sum of averages of the summands. The same is true for the variance of random sum. These relationship hold for more than two independent random variables. Let us write these relationships mathematically: the random sum $\sum_i Y_i$ of independent random variables has the average value

$$\left\langle \sum_i Y_i \right\rangle = \sum_i \langle Y_i \rangle \ ,$$

and variance

$$\left\langle \delta \left(\sum_i Y_i \right)^2 \right\rangle = \sum_i \langle \delta Y_i^2 \rangle \ .$$

3.3.7 Random Process

It is very important to remember that our earlier probability assignment in (3.1) was based on the knowledge about the lot: we assume that n of the total are in that form, and that n is a fixed known number. This should not be confused with the fraction $x(t) = n(t)/n^{\text{tot}}$ of molecules that are unmodified, defined in the last chapter in a deterministic setting. Therein, $n(t)$ represented a deterministic time-dependent protein abundance, which, in the stochastic setting, has been replaced by the random process $N(t)$. The deterministic time-dependent fraction $x(t)$ satisfies the differential equation (2.21):

$$\dot{x}(t) = k_u - (k_w + k_u) x(t) \ .$$

How can we relate the deterministic quantities $n(t)$ and $x(t)$ to the random process $N(t)$? Before we proceed with that, recall that, unlike a unique time function $n(t)$, the random process $N(t)$, as a rule, will assign a different time function to each time-course measurement made for the abundance. The average features of the process will appear in all the time functions, but in addition, each time function will have features specific to that measurement and representing all the factors that were, willingly or unwillingly, excluded from the model. In light of this, the only sensible way in which we can relate the deterministic time function to the random process is to suggest (or hope) that the former $n(t)$ is the average feature of the latter and hence should be reflected in all the time functions assigned to the time-course measurements by the process $N(t)$. Denoting the average time-dependent protein abundance by $\langle N(t) \rangle$, we suggest that the following relation holds:

$$\langle N(t) \rangle = n(t) = n^{\text{tot}} x(t).$$

Combining this with (3.4) gives

$$p_u(t) = \frac{\langle N(t) \rangle}{n^{\text{tot}}} = x(t).$$

Thus, we can replace $x(t)$ in (2.21) with $p_u(t)$ to get the kinetic law for the success probability,

$$\dot{p}_u = k_u - (k_w + k_u) p_u.$$

The steady-state solution, obtained by equating the derivative to zero, is

$$p_u^{\text{ss}} = \frac{\langle N^{\text{ss}} \rangle}{n^{\text{tot}}} = \frac{k_u}{k_w + k_u}. \tag{3.5}$$

If the protein molecule is initially modified, that is, $p_u(0) = 0$, the time-dependent solution is

$$p_u(t) = \frac{k_u}{k_w + k_u} \left[1 - e^{-(k_w + k_u)t} \right]. \tag{3.6}$$

The steady-state solution (3.5) can be interpreted as the fraction of time spent by the protein in the unmodified form, which makes intuitive sense in a chemical equilibrium.

3.3.8 Probability Distribution

Let us look at the experiment of selecting molecules from another angle. Although we do not know the proportions of molecules in the two forms, we know the probability $p_u(t)$ that an individual protein molecule is unmodified. How can we use this information to say something about the abundance of unmodified proteins? Extend the experiment of selecting two molecules to selecting all the n^{tot} molecules in a sequence with replacement (each molecule is placed back after having been selected). The form of the protein molecule in the ith selection, in a sequence, is represented by a random variable Y_i that assigns to the selection a value (1)0 for the (un)modified form. Then the random variable $N(t) = \sum_{i=1}^{n^{\text{tot}}} Y_i$ represents the number of molecules that are unmodified. The probability of each selection to be in one of the two forms is the same because of independence of events. Consider the specific sequence that has $Y_i = 1$ for the first n selections and $Y_i = 0$ for the rest of

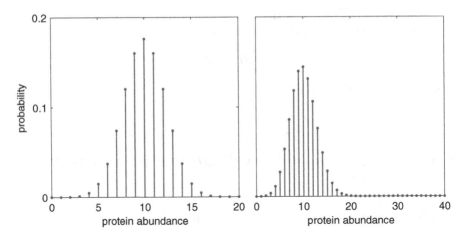

Figure 3.1 Steady-state probability distribution of protein abundance for the standard modification. The distribution is binomial with parameters n^{tot} and p_u with the average located at $n^{\text{tot}}p_u = 10$. Left: $n^{\text{tot}} = 20$, $p_u = 0.5$, symmetric distribution around the average. Right: $n^{\text{tot}} = 40$, $p_u = 0.25$, distribution is skewed toward smaller values.

the sequence. The associated per-selection probabilities are

$$\Pr[Y_i = y] = \begin{cases} p_u & \text{if } y = 1,\ 1 \leq i \leq n, \\ 1 - p_u & \text{if } y = 0,\ n < i \leq n^{\text{tot}}. \end{cases}$$

Now the probability of this specific sequence is, following the multiplication rule for independent events,

$$\Pr\left[\bigcap_{i=1}^{n} Y_i = 1 \bigcap_{k=n+1}^{n^{\text{tot}}} Y_k = 0\right] = \prod_{i=1}^{n} \Pr[Y_i = 1] \prod_{k=n+1}^{n^{\text{tot}}} \Pr[Y_k = 0]$$

$$= p_u^n \left(1 - p_u\right)^{n^{\text{tot}} - n}.$$

The specific sequence is one of

$$\binom{n^{\text{tot}}}{n} = \frac{n^{\text{tot}}!}{n!\,(n^{\text{tot}} - n)!}$$

distinct sequences that have n unmodified protein molecules, and with the same probability. Since all the sequences are disjoint, their sum

$$\Pr[N(t) = n] = \binom{n^{\text{tot}}}{n} \left(p_u(t)\right)^n \left(1 - p_u(t)\right)^{n^{\text{tot}} - n} \tag{3.7}$$

gives the probability that n molecules in the lot, not just in one specific sequence, are unmodified proteins. The above equation expresses a *binomial distribution* with parameters n^{tot} and p_u. The distribution has average value $\langle N(t) \rangle = n^{\text{tot}} p_u(t)$ and variance $\langle \delta N^2 \rangle = n^{\text{tot}} p_u (1 - p_u)$, where the latter follows from the fact that the sum of independent random variables has a variance that is the sum of their individual variances. The distribution is plotted in Figure 3.1 for selected values of parameters n^{tot} and p_u. When viewed in steady state, this also highlights another important difference between the two frameworks. In the deterministic framework, we say that the protein abundances in the two forms have reached constant values $n^{\text{ss}} = n^{\text{tot}} p$ and $n^{\text{tot}} - n^{\text{ss}}$. In the stochastic framework, we say that the probability distribution has reached a constant distribution with the average abundances $\langle N^{\text{ss}} \rangle = n^{\text{tot}} p_u^{\text{ss}}$ of unmodified protein molecules. In other words, the deterministic steady state appears as the average value of the stochastic steady-state distribution.

Initial conditions: The probability distribution described by (3.7) implicitly assumes some fixed initial condition $N(0)$. If all the molecules are initially in the modified form, that is, $N(0) = 0$, then $p_u(0) = 0$, in which case, the time-dependent success probability $p_u(t)$ is given by (3.6).

3.3.9 Measures of Fluctuation

For a random process $N(t)$, the average time function $\langle N(t) \rangle$ is a measure of the expected time course of a (time-dependent) distribution, and the variance $\langle \delta N^2 \rangle$ is a measure of the spread of the distribution around the average. This should give the impression that the variance can be used as a measure of fluctuations. That is not an appropriate choice because of different units (due to the squaring involved in the variance). The fluctuation around the average can then be measured by the square root $\sqrt{\langle \delta N^2 \rangle}$ of the variance. Fluctuation so defined is also known as standard deviation (SD). Even this may be inappropriate when two fluctuations around averages at different scales need to be compared, because the scale of each fluctuation would inherit the scale of the corresponding average. Therefore, a more appropriate measure of noise is the relative fluctuation around the average, measured as the SD divided by the average, also known as the coefficient of variation (CV). The relative fluctuation for the protein abundance process $N(t)$ above is

$$\frac{\sqrt{\langle \delta N^2 \rangle}}{\langle N \rangle} = \frac{\sqrt{n^{\text{tot}} p_u (1 - p_u)}}{n^{\text{tot}} p_u} \propto \frac{1}{\sqrt{n^{\text{tot}}}},$$

and thus inversely proportional to the square root of the system size, which is a general rule of thumb. The implication here is that stochasticity due to discreteness is not significant for large abundances but has to be taken into account for small abundances. More will be said on measures of fluctuation in the coming chapters. To get a glimpse of how different choices of the measure lead to different consequences, have a look at Figures 5.13 and 6.1.

3.3.10 Rare Events

The binomial distribution is skewed toward the smaller protein abundances when n^{tot} is doubled and p_u halved so that the average stays the same. If we keep increasing n^{tot} with a proportional decrease in p_u so that the average $n^{\text{tot}} p_u$ does not change, the distribution will be skewed more and more toward the left. To interpret this physically, it follows from (3.6) that smaller values of p_u can be achieved by setting k_u smaller compared to k_w. That in turn, for a single molecule, is equivalent to fast modification and slower demodification. A proportional increase in n^{tot} will make up for the slower demodifications to keep the average $n^{\text{tot}} p_u$ the same. Let us investigate the limiting behavior of the above (binomial) distribution for infinitely large n^{tot} and infinitely small p_u such that their product is a finite positive number, $n^{\text{tot}} p_u = \mu$. We start by noting that

$$\binom{n^{\text{tot}}}{n} p_u^n = p_u^n \frac{n^{\text{tot}} (n^{\text{tot}} - 1) \cdots (n^{\text{tot}} - n + 1)}{n!}$$

$$= \frac{(n^{\text{tot}} p_u)^n}{n!} \left(1 - \frac{1}{n^{\text{tot}}}\right) \cdots \left(1 - \frac{n-1}{n^{\text{tot}}}\right)$$

$$\rightarrow \frac{\mu^n}{n!} \quad \text{as } n^{\text{tot}} \rightarrow \infty .$$

Similarly,

$$(1 - p)^{n^{\text{tot}}} = \left(1 - \frac{\mu}{n^{\text{tot}}}\right)^{n^{\text{tot}}} \rightarrow e^{-\mu} \text{ as } n^{\text{tot}} \rightarrow \infty,$$

and $(1 - p)^{-n} \rightarrow 1$ as $p \rightarrow 0$. Thus the binomial distribution (3.1) approaches a distribution with one parameter λ, namely

$$\Pr\left[N(t) = n\right] = \frac{e^{-\mu(t)} (\mu(t))^n}{n!} . \tag{3.8}$$

This is called the *Poisson distribution* with average value $\mu(t)$. An infinitely small p_u also means that the variance is the same as the average value μ. This is a very important and useful observation: when you access to the average value and variance of an otherwise unknown distribution, you look at the value of the variance relative to the average, and the ratio will tell you how close/faraway the distribution is to/from the Poisson distribution. Since a small p_u means a *rare* event, this last result is known as the *law of rare events*. What is essentially meant here is that when an extremely large number of extremely rare events are possible during some time interval, the number of events occurring during the interval is a Poisson distributed random variable with a finite average value. The situation is easily understood in terms of calls received by a telephone exchange operator. Although calls from individual callers are rare in a large population, the number of calls received by the operator during a time interval is not necessarily small.

3.4 The Poisson Process

Let us dig further into the dynamics (temporal aspects) of rare events. In the limiting procedure leading to the Poisson distribution above, the two parameters n^{tot} and $p_u(t)$ disappeared and a new parameter $\mu(t) = n^{\text{tot}} p_u(t)$ emerged to replace them. Therefore, for pedagogical reasons, we shall not use the standard (de)modification for investigating the dynamics of rare events. Instead, we select a zero-order reaction

$$\varnothing \xrightarrow{\ \lambda\ } X$$

in which molecules of some protein (the only species) X are produced at a constant rate from some reactant that stays constant during the time scale of interest (because of infinite abundance or constant replenishment). Note that $\lambda = \hat{k}$ is the conversion rate constant. In a deterministic setting, the time-dependent protein abundance $n(t)$ has kinetics $\dot{n}(t) = \lambda$, which has a solution $n(t) = \lambda t$ that increases linearly with time from the supposed zero initial abundance. In a stochastic setting, this time function can be interpreted as the average trajectory $\langle N(t) \rangle = \lambda t$ of the random process $N(t)$ that represents the time-dependent abundance. We are interested in the random process $N(w+t) - N(w)$ representing the abundance increment during a time interval $[w, w+t]$ with known $N(w)$. The average increment during the interval is

$$\langle N(w+t) - N(w) \rangle = \lambda(w+t) - \lambda w = \lambda t$$

which does not depend on the time w of measurement. If we divide the interval into a large number K of subintervals, each of length $\Delta t = t/K$, so short that at most one reaction could occur in a subinterval with probability $\lambda \Delta t$, then the increment to the average abundance during a subinterval is

$$\langle \Delta N \rangle = \lambda \Delta t = \frac{\lambda t}{K}.$$

The increment ΔN is a Bernoulli random variable because of only two possible events: a success $\Delta N = 1$ (one reaction) with probability $p = \lambda \Delta t$ and a failure $\Delta N = 0$ (no reaction) during the subinterval. One point to note here is that since $\lambda \Delta t$ is the probability of one reaction instance during the short interval, the conversion rate λ can be interpreted as the probability per unit time of the occurrence of the reaction. We note that the success probability $p = \lambda \Delta t$ depends neither on the time w of measurement nor on the associated state $N(w)$. The random process can then be expressed as

$$N(w + t) - N(t) = K \Delta N(t),$$

which is a sum of K Bernoulli variables, and hence has a binomial distribution with parameters K and p. Since we require a very large K to keep the subinterval very short, resulting in a very small p to give rare events, the binomial distribution approaches a Poisson distribution,

$$\Pr[N(w + t) - N(w) = n] = \Pr[N(t) = n] = \frac{e^{-\lambda t}(\lambda t)^n}{n!}.$$

with the average and variance both equal to $\mu(t) = \lambda t$. The random process with the above time-dependent probability distribution is called the *Poisson process* with rate parameter λ. No dependence on w means that the increments in nonoverlapping intervals are stationary. No dependence on $N(w)$ means that they are identically distributed.

3.4.1 Interreaction Time

The first interesting thing to be observed is that the above probability reduces to $e^{-\lambda t}$ for $n = 0$, which means that the time T until the next reaction, that is, the times between successive reactions, has probability $e^{-\lambda t}$ of being greater than t. Mathematically, this is expressed by

$$\Pr[T > t] = e^{-\lambda t}.$$

The continuous random variable T_0 with the above property is said to be exponentially distributed, or to be an exponential random variable, with parameter

λ. Note that the distribution does not depend on w or $N(w)$. In other words, the interreaction times are independent and identically distributed random variables with a common exponential distribution.

Exponential density: We have seen that the time T_n until the next reaction is exponentially distributed, based on the expression $e^{-\lambda t}$ for the probability $\Pr[T_n > t]$ that the occurrence of the next reaction is later than t units of time. The probability $\Pr[T_n \le t]$ that the reaction occurs within the next t units of time is, complimentarily, $1 - e^{-\lambda t}$. Since this probability can be expressed as the integral

$$1 - e^{-\lambda t} = \Pr[T_n \le t] = \int_{-\infty}^{t} p(x)\mathrm{d}x$$

of the probability density, the latter is simply the time derivative of the probability, that is,

$$p(t) = \lambda e^{-\lambda t},$$

the *exponential density*.

3.4.2 Arrival Times

Apart from interreaction times, we may also be interested in the time W_n of the nth reaction. This is the time the process must wait before state n is reached, and is referred to as the *arrival time* or *epoch time* of the nth reaction. Its probability density can be worked out by considering the possibility of the nth reaction taking place during the interval $[t, t + \Delta t]$. This is equivalent to the statement that the first $n - 1$ reactions have occurred in the interval $[0, t]$ and one reaction occurs in the next interval of length Δt. So the event of our interest is a joint event involving two intervals. Since they are nonoverlapping, the number of events that occur within one interval is independent of the number of events that occur within the other interval. Thus the required probability of the joint event is

$$\begin{aligned}
\Pr[t \le W_n \le t + \Delta t] &= \Pr[(N(t) = n - 1) \cap (\Delta N(t) = 1)] \\
&= \Pr[N(t) = n - 1]\Pr[\Delta N(t) = 1] \\
&= \frac{e^{-\lambda t}(\lambda t)^{n-1}}{(n-1)!}\lambda\Delta t,
\end{aligned}$$

and hence the probability density of the nth arrival time W_n is

$$p(t) = \frac{\lambda^n t^{n-1}}{(n-1)!} e^{-\lambda t}, \tag{3.9}$$

which expresses an *Erlang distribution* with rate parameter λ and shape parameter n.

Exponential random sums: Since the arrival time can be expressed as a sum $W_n = \sum_{i=0}^{n-1} T_i$ of n interarrival times, all exponentially distributed with the same parameter λ, we conclude that the sum of exponential random variables is an Erlang random variable with shape parameter that is the number of summands and rate parameter that is the common parameter of the exponential distributions. A detailed discussion of the Erlang distribution and its application will be presented in the next chapter, in Section 4.5.2.

3.5 Chemical Reactions as Events

In the last section we noted how the Poisson process was a natural choice to describe the time-dependence abundance $N(t)$ of the zero-order reaction

$$\varnothing \xrightarrow{\ \lambda\ } X,$$

with the important discovery that the conversion rate λ, appearing as the rate parameter of the Poisson process $N(t)$, has a new interpretation as the probability per unit time of the occurrence of the reaction. We now would like to investigate whether a similar interpretation can be given to the conversion rate of a general elementary reaction.

3.5.1 Conversion Rate

Consider again the standard (de)modification (2.20). The two conversion rates are

$$\dot{z}_w(t) = k_w n(t), \quad \dot{z}_u = k_u \left(n^{\text{tot}} - n(t)\right).$$

The reaction-count increments during an infinitesimally short time interval $[t, t + \Delta t]$ are

$$\Delta z_w(t) = k_w n(t)\Delta t, \quad \Delta z_u(t) = k_u \left(n^{\text{tot}} - n(t)\right).$$

It is important to realize that $n(t)$, $z_w(t)$, and $z_u(t)$ are deterministic time functions. In a stochastic setting, these will be replaced by stochastic processes $N(t)$, $Z_w(t)$, and $Z_u(t)$. If n represents a sample state of $N(t)$, then the above reaction-count increments can be interpreted only in an average sense:

$$\langle \Delta Z_w \rangle_n = k_w n \Delta t, \quad \langle \Delta Z_u \rangle_n = k_u \left(n^{\text{tot}} - n \right) \Delta t .$$

The subscript n here reflects the conditional average such that the probability in (3.2) is conditioned on state n. Both the reaction-count increments ΔZ_w and ΔZ_u are Bernoulli variables, since at most, one reaction can occur in the infinitesimal interval. Since the average of a Bernoulli variable is equal to its success probability, the expressions in the above equations are (success) probabilities of one reaction of the corresponding type. It then follows that the conversion rates $k_w n$ and $k_u \left(n^{\text{tot}} - n \right)$ are probabilities per unit time of the occurrence of the modification and demodification, respectively. The important difference between this and the Poisson process is the dependence of these probabilities on the current state n. However, the process has no memory, since the probabilities of future events depend only on the current state, and not on past states. Such a process is said to be memoryless or to have the Markov property. In other words, we have a Markov process.

We can follow a similar analysis for any elementary reaction and thus conclude that *the conversion rate of an elementary reaction is the probability per unit time of its occurrence*. In Chapter 5, this will be referred to as the *reaction propensity*.

3.5.2 Interreaction Time

Suppose the copy-number process is in state n. Write $T_w(n)$ for the time until the next modification reaction and $T_u(n)$ for that until the next demodification reaction. The probability that a modification occurs in the next Δt time units is $k_w n \Delta t$. As long as n does not change, the occurrence of the next modification reaction is a Poisson arrival, the time until which, $T_w(n)$, is an exponential random variable with parameter $k_w n$. Similarly, the time $T_u(n)$ until the next demodification reaction is an exponential random variable with parameter $k_u n$. If we could somehow generate random numbers from an exponential distribution, then it is straightforward to generate sample paths of the stochastic process $N(t)$. Simply generate two random numbers t_w and t_u as samples of the exponentially distributed random variables $T_w(n)$ and $T_u(n)$, respectively. The smaller of the two will determine the time and type of the next reaction. Alternatively, we could work with the time $T_0(n)$ until the next reaction (of any type). The probability of one reaction, irrespective of its type, during the next Δt time units is $(k_w + k_u) n \Delta t$. As long as n does not change, the occurrence of the next reaction is a Poisson arrival, the time until which, $T_0(n)$, is an exponential random variable with parameter $(k_w + k_u) n$. Thus we need to generate only one random number t_0 as a sample of the exponentially distributed random variable $T_0(n)$. The type of the next reaction, known to have occurred, is modification with probability $k_w/(k_w+k_u)$ and demodification with probability $k_u/(k_w+k_u)$. These expressions follow from the fact that the probabilities are conditional. This is the key idea behind the stochastic simulation method that will be formally developed in Chapter 5.

So far we have presented the notions of probability, random variables, and stochastic processes, using intuition without the mathematical theory of probability. The next chapter follows a formal and rigorous discourse and hence should complement this chapter.

Problems

3.1. Transcription factors bind to a promoter region to regulate gene expression. Suppose three repressors and one activator can bind a promoter region in a sequential manner. Imagine the observations of a sequence of molecules binding to the promoter until the molecule turns out to be an activator. What is the probability that we find an activator on the seventh observation?

3.2. Two events A and B are said to be mutually exclusive (or disjoint) when the occurrence of one excludes the possibility of the occurrence of the other. Imagine a stem cell dividing into two daughter cells, with only two possibilities: both daughter cells are identical, such an event being denoted by A, or they are different, the event being denoted by B. What is the mathematical representation of the fact that A and B are mutually exclusive? What is the conditional probability of one given the other?

3.3. Two events A and B are said to be independent when the occurrence of one does not change the probability of occurrence of the other. Consider two stem cells, each dividing into two daughter cells, again with only two possibilities. Let us write A_1 for the event that the first stem cell divides into identical cells, and B_1 for the opposite. The events A_2 and B_2 can be similarly defined for the second stem cell. Assume that the two cells do not influence each other and are not influenced by a common external agent. Identify, among the four event, pairs of

1. mutually exclusive events.

2. independent events.

3.4. Consider two stem cells that rarely die and do so independently of each other. Assume a probability 10^{-6} of the rare death of each stem cell. What is the probability that they both die? What is the probability that at least one of the cells dies?

Chapter 4

Probability and Random Variables

The previous chapter provided an informal introduction to concepts that are necessary for stochastic modeling. So the readers who are happy with that material may skip this chapter on first reading and return to it when needed in the later chapters. It has been added as background necessary for an advanced understanding of the subsequent chapters. The books by Breuer and Petruccione [18] and Allen [5] were the main inspirations for this chapter.

4.1 Probability: A Measure

The central notion of probability theory is a random experiment: certain conditions are fixed, and the observed phenomena are recorded to give the outcome of the experiment. The fixed conditions cannot uniquely determine the outcome of the experiment, which is influenced by other conditions (known or unknown) not fixed. These conditions are explained as the effect of chance (or randomness). Hence the explanation of the notion of *chance* (randomness) differs essentially from its semantic content in everyday life. There is no sense of rareness and unexpectedness, but it is a simple expression of ignorance about what will influence and determine the observed outcome of a future experiment.

We introduce the terminology used in probability theory in terms of sets. Each possible *outcome* of an experiment is considered to be a sample point, or an element ω, of a set Ω, the *sample space*. An outcome of interest is expressed by the notion of an *event* defined as a set of sample points, which is some subset of the sample space. To understand the meaning of an event, we say that an event A has occurred if the outcome is an element of A. The sample space Ω, being a collection of all possible outcomes, is the certain event, and the null set $\emptyset$ is the *impossible event*. Subsets $\{\omega\}$ containing just one element of are called *elementary events*. The usual operations with sets can now be interpreted in the language of the corresponding events:

- The *joint occurrence* of (both the) two events A_1 and A_2 is represented by their intersection $A_1 \cap A_2$.

- The *alternative occurrence* of (at least one of) the two events is represented by their union $A_1 \cup A_2$.

- The *nonoccurrence* of the event A is represented by its complement $\Omega \setminus A$.

- The set inclusion $A_1 \subset A_2$ represents the *implied occurrence* of A_2 whenever A_1 occurs.

- The set difference $A_1 \setminus A_2$ represents the *implied nonoccurrence* of A_2 whenever A_1 occurs.

This way we can represent mathematically interesting situations between events. Probability is then defined as a measure of chance assigned to an event. All subsets of Ω may not be of interest to us (for example, if Ω is uncountable). All possible events of interest are expressed by a collection $\mathcal{A}$ of events including:

1. the sample space Ω itself and the empty set $\emptyset$,

2. the union $A_1 \cup A_2$, the intersection $A_1 \cap A_2$, and the difference $A_1 \setminus A_2$ of any two events $A_1, A_2 \in \mathcal{A}$, and

3. the union $A_1 \cup A_2 \cup \cdots$ of any countable event collection $A_1, A_2, \ldots \in \mathcal{A}$.

Such a collection $\mathcal{A}$ is our event space, which contains all the events of interest. We can now introduce probability as a measure of chance normalized by the chance of the certain event. The impossible event has no chance; hence its probability is zero. The certain event has probability equal to unity. Any other event is assigned a real number between 0 and 1 that quantifies its chance of occurrence. Formally, a *probability measure* is a map $\Pr : \mathcal{A} \to [0,1]$ that assigns to each event A a real number $\Pr[A]$ in the unit interval, the assignment being denoted by $A \mapsto \Pr[A]$. The number $\Pr[A]$ is a measure of the size of A relative to Ω and is interpreted as the probability of the event A. The probability measure $\Pr$ must satisfy the following additive rule, the Kolmogorov axiom: If we have a countable collection of disjoint events

$$A_1, A_2, \ldots \in \mathcal{A}, \quad \text{with } A_i \cap A_j = \emptyset \text{ for } i \neq j,$$

then the probability of their union is equal to the sum of their probabilities,

$$\Pr[A_1 \cup A_2 \cup \cdots] = \Pr[A_1] + \Pr[A_2] + \cdots .$$

It must be possible to assign a probability, according to the above requirements, to every event in the event space $\mathcal{A}$. Formally speaking, the events in $\mathcal{A}$ are said to be *measurable events*.

On the basis of these axioms one can build up a consistent theory of probabilities. In particular, the Kolmogorov axiom enables one to determine the probabilities for all events that arise from logical operations on other events. For example, one finds that

$$\Pr[A_1 \cup A_2] = \Pr[A_1] + \Pr[A_2] - \Pr[A_1 \cap A_2] . \tag{4.1}$$

4.1.1 Probability Space

We have been talking about two types of spaces: the sample space Ω of all possible outcomes of our experiment and the event space $\mathcal{A}$ of all events of interest to which probabilities can be assigned. The notion of random experiment can be characterized by these two spaces together with the probability measure. The triplet $(\Omega, \mathcal{A}, \Pr[\cdot])$ thus formed is called a *probability space*. This concept of a probability space constitutes the axiomatic basis of classical probability theory. Of course, from a physical viewpoint one has to relate these abstract notions to an experimental setting. We can also look at probability as a measure of our *uncertainty* about the event. Thus if we are almost certain that an event will occur, we assign to it a probability close to 1, say 0.99. We assign a number close to zero, say 0.01, if we are almost certain that it will not occur. When our uncertainty about the event is maximum, we assign 0.5.

4.1.2 Conditional Probability

In most applications, there exists information that when taken into account, alters the assignment of probability to events of interest. This concept is often formulated by introducing the *conditional probability* $\Pr[A_1|A_2]$ of an event A_1 under the condition that an event A_2 has occurred,

$$\Pr[A_1|A_2] = \frac{\Pr[A_1 \cap A_2]}{\Pr[A_2]}, \tag{4.2}$$

where the denominator accounts for the reduced sample space after the occurrence of event A_2. Of course, both events A_1 and A_2 are taken from the event space $\mathcal{A}$, and it is assumed that $\Pr[A_2] > 0$. These events are said to be *independent* if

$$\Pr[A_1|A_2] = \Pr[A_1],$$

or equivalently, if

$$\Pr[A_1 \cap A_2] = \Pr[A_1]\Pr[A_2] .$$

This means that the probability of the joint occurrence of two independent events factor to their individual probabilities. For more than two events the condition of mutual independence means that the probability of the occurrence of any subset of joint events factors as above. It is important to note that mutual independence is not always implied by pairwise independence.

If a collection $B_1, B_2 \ldots$, of pairwise disjoint nonempty events *partitions* the sample space, that is, $\Omega = \cup_i B_i$, then combining the third axiom of probability with (4.2), we can write for an event A,

$$\Pr[A] = \sum_i \Pr[B_i] \Pr[A|B_i] . \tag{4.3}$$

This law, called the "total law of probability", can be extended by conditioning all events on a further event C:

$$\Pr[A|C] = \sum_i \Pr[B_i] \Pr[A|B_i|C] .$$

4.2 Random Variables

The elements ω of the sample space Ω can be rather abstract objects. In practice, one often deals with simple numbers (integer or real). For example, one would like to add and multiply these numbers, and also to consider arbitrary functions of them. The aim is thus to associate real numbers with the elements of the sample space. This idea leads to the concept of a *random variable*.

Consider a probability space $(\Omega, \mathcal{A}, \Pr[\cdot])$. A random variable may be defined as a mapping that assigns a real number to each element of the sample space Ω. Thus every event in $\mathcal{A}$ will map to a corresponding subset of $\mathbb{R}$. Since we would like to deal with real numbers, every subset (of $\mathbb{R}$) of interest must also must map back to an event in $\mathcal{A}$, under the inverse mapping. Thus every subset B of interest must have a corresponding event A in $\mathcal{A}$ (under the inverse map). Formally, such subsets are called *Borel sets*. Do not be intimidated by this term: a Borel set is a subset of $\mathbb{R}$ that has a corresponding measurable event in the event space $\mathcal{A}$, where the adjective *measurable* indicates the assignability of probability. This puts a constraint on the choice of functions available for defining a random variable. Now we can give a formal definition of a random variable.

For a probability space $(\Omega, \mathcal{A}, \Pr[\cdot])$, a random variable X is defined to be a real-valued map

$$X : \Omega \to \mathbb{R} \tag{4.4}$$

that assigns to each elementary event $\omega \in \Omega$ a real number $X(\omega)$ such that

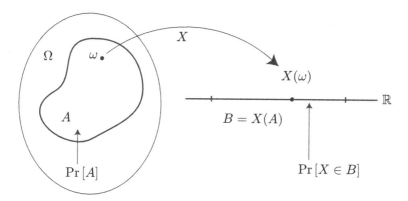

Figure 4.1 Illustration of the definition of a random variable. A random variable X is a map from the sample space S to the real line $\mathbb{R}$. Probability is assigned to a Borel set B using $\Pr[X \in B] = \Pr[A]$, where $A = X^{-1}(B)$. Figure adopted from [18].

every Borel set B has a corresponding event,

$$X^{-1}(B) = \{\omega \in \Omega : X(w) \in B\},$$

that belongs to the event space $\mathcal{A}$. The event $X^{-1}(B)$, corresponding to the Borel set B, will be written as $X \in B$ for convenience.[1] Thus we will write $X = x$ for the event corresponding to the singleton $\{x\}$, $x_1 < X \leq x_2$ for the one corresponding to $(x_1, x_2]$, $X \leq x$ for the one corresponding to $(-\infty, x]$, and so forth. This notation is only a matter of convenience, and the reader should always remember that X is a mapping, not a number. The above condition ensures that probabilities can be assigned to events $X \in B$ as illustrated in Figure 4.1. Given some ω, the value $x = X(\omega)$ is a sample, or *realization*, of X. In the following we use the usual convention to denote a random variable by a capital letter, and its realization by the corresponding lowercase letter. The range set $X(\Omega)$, the set of all realizations of X, is the *state space* of X. If the state space $X(\Omega)$ is finite or countably infinite, then X is a *discrete random variable*. If the state space $X(\Omega)$ is uncountable (typically an interval), then X is a *continuous random variable*. However, the random variable could be of *mixed type*, having properties of both discrete and continuous random variables. The probability that a random variable X takes a value less than or equal to a given value x,

$$F_X(x) = \Pr[X \leq x],$$

---

[1]We will not use the usual, but misleading, notation $\{X \in B\}$, which may be misunderstood as a set with element $X \in B$.

is the most general way to characterize X. The function $F_X(\cdot)$ is the *cumulative distribution function* (CDF), or simply the *distribution function* of X. The subscript X, implicit from the context, will often be dropped. Sometimes, it is more useful to work with a closely related *right-tail distribution function*, or *complementary cumulative distribution function* (CCDF):

$$G(x) = \Pr[X > x] = 1 - F(x).$$

Thus the distribution function $F(x)$ is the probability that X falls to the left of the point x (on the real line), and $G(x)$ gives the probability for X falling to the right of x.

4.2.1 Discrete Random Variables

For a discrete random variable X taking values n from some state space, the CDF $F(n)$ is a (discontinuous) step function that can be written as

$$F(n) = \sum_{n' \leq n} P(n'),$$

where $P(\cdot)$ is the *probability mass function* (PMF) defined for each sample n by the probability

$$P(n) = \Pr[X = n].$$

The PMF can be used to compute the probability that X falls in a (Borel) set B,

$$\Pr[X \in B] = \sum_{n \in B} P(n).$$

In the following we give a few familiar and important examples of discrete random variables and their distributions.

Bernoulli distribution: A Bernoulli variable X has two possible values: $x = 1$, a success, with probability p, and $x = 0$, a failure, with probability $1 - p$. The probability distribution with parameter p can be written as

$$\Pr[X = x] = p^x (1 - p)^{1-x}, \quad x = 0, 1.$$

We write $X \sim \text{Bernoulli}(p)$ to declare X as a Bernoulli random variable with success probability p. The PMF above is zero for values of x other than 0 and 1. This needs to be understood whenever a probability distribution is expressed in terms of PMF or PDF:

The PMF/PDF is zero for sample values other than those specified in the formula expressing the distribution.

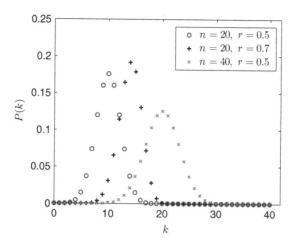

Figure 4.2 Binomial distribution for selected parameter values.

Binomial distribution: The number X of successes in n independent trials, with probability p of success at each trial, takes integer values k with probabilities

$$\Pr[X = k] = \binom{n}{k} p^n (1 - p)^{n-k}, \quad 0 \leq k \leq n. \tag{4.5}$$

This set of probabilities forms the *binomial distribution* with parameters n and k, and we write $X \sim \text{Binomial}(n, k)$ to represent that X follows this distribution. It can be seen that the binomial random variable is, in essence, the sum of n Bernoulli variables with a common parameter p. The form of the distribution follows by considering that $p^k(1 - p)^{n-k}$ gives the probability of one particular sequence of k successes (and $n - k$ failures) and $\binom{n}{k}$ is the number of different sequences possible. Figure 4.2 shows the PMF of the binomial distribution for three different pairs of parameter values.

Poisson distribution: A discrete random variable X with probabilities

$$\Pr[X = n] = \frac{e^{-\lambda}\lambda^n}{n!}, \quad n = 0, 1, 2, \ldots, \tag{4.6}$$

is said to have the *Poisson distribution* with parameter λ and is denoted by $X \sim \text{Poisson}(\lambda)$. In our discussion leading to (3.8), we saw how the binomial distribution approaches the Poisson distribution when there is a large number of trials, each with rare success. Figure 4.3 shows the PMF of the Poisson

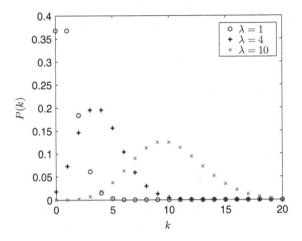

Figure 4.3 Poisson distribution: PMF for selected parameter values.

distribution for selected values of the rate parameter λ.

Geometric distribution   The number X of trials until the first success in a sequence of trials with probability p of success at each trial has the *geometric distribution*

$$\Pr\left[X = n\right] = (1 - p)^n p, \qquad n = 0, 1, 2, \ldots,$$

with parameter p, and we write $X \sim \text{Geometric}\,(p)$ to represent that X follows this distribution. Figure 4.4 shows the PMF of the geometric distribution for three selected values of the parameter p.

4.2.2 Continuous Random Variables

For a continuous random variable X taking values x from some state space, the CDF $F(x)$ is continuous (from the right) and can be written as

$$F(x) = \int_{-\infty}^{x} p(x')\mathrm{d}x',$$

where $p(x)$ is the *probability density function* (PDF), or simply the density function of X. From the definition above, it follows that

$$p(x) = \frac{\mathrm{d}F}{\mathrm{d}x}$$

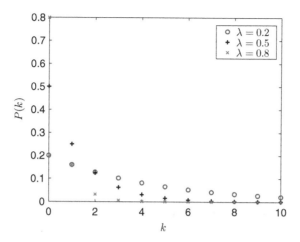

Figure 4.4 Geometric distribution: PMF for selected parameter values.

for regions (of the state space of X) where F is differentiable. Loosely speaking, $p(x)$ can be interpreted such that,

$$\Pr\left[x \leq X < x + h\right] \approx p(x)h, \quad \text{for small } h,$$

which justifies that $p(x)$ is a density. The probability that X falls in a Borel set B is then given by

$$\Pr\left[X \in B\right] = \int_B p(x)\mathrm{d}x \,.$$

A few important examples of continuous random variables are given next.

Uniform distribution: The position X of a point selected at random in some interval $[a, b]$ such that all subintervals of equal lengths are equiprobable has the PDF

$$p(x) = \frac{1}{b - a}, \ a \leq x \leq b\,.$$

This is the density of the so-called *uniform distribution*. That X follows the uniform distribution in the interval $[a, b]$ is written as $X \sim \text{Uniform}\,(a, b)$. The CDF of X can be obtained by integrating the above density:

$$F(x) = \Pr\left[X \leq x\right] = \frac{x - a}{b - a}, \ a \leq x \leq b\,.$$

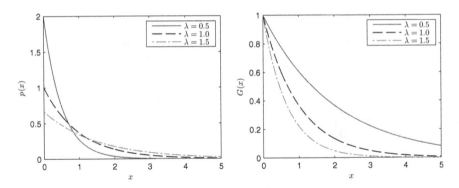

Figure 4.5 Exponential distribution. *left*: The PDF for three different values of the rate parameter λ. *right*: The CCDF for the same parameter values.

The CDF is zero/unity for values of x to the left/right of the interval $[a, b]$. This needs to be understood whenever a probability distribution is expressed in terms of CDF:

> *The CDF is zero/unity for sample values lower/higher than those specified in the formula expressing the distribution.*

Of particular interest is the standard uniform variable $U \sim \text{Uniform}(0, 1)$ with the distribution function

$$\Pr\left[U \leq u\right] = \frac{u - 0}{1 - 0} = u, \ 0 \leq u \leq 1. \tag{4.7}$$

The standard uniform random variable will turn out to be useful in generating samples of a random variable with an arbitrary distribution.

Exponential distribution: A random variable X is said to be *exponential* with parameter λ, denoted by $X \sim \text{Exponential}(\lambda)$, if its PDF is the exponential density,

$$p(x) = \lambda e^{-\lambda x}, \ x \geq 0. \tag{4.8}$$

The *time* until an event from a sequence of events, the occurrences of which over nonoverlapping intervals are independent, such as Poisson arrivals, is exponentially distributed. The CCDF of X is then given by

$$G(x) = \Pr\left[X > x\right] = \int_{x}^{\infty} p(x)\mathrm{d}x = e^{-\lambda x},$$

and the CDF, by $F(x) = 1 - G(x) = 1 - e^{-\lambda x}$. Figure 4.5 illustrates how the exponential distribution looks for three different values of the rate parameter, where the PDF $p(x)$ is plotted to the left and the CCDF $G(x)$ to the right.

4.2.3 The Memoryless Property

The exponential distribution has the remarkable property that it does not remember its past. To see this, consider the (conditional) probability

$$\Pr\left[X > t + s \mid X > t\right] = \frac{\Pr\left[X > t + s \cap X > t\right]}{\Pr\left[X > t\right]} = \frac{\Pr\left[X > t + s\right]}{\Pr\left[X > t\right]}$$
$$= \frac{G(t + s)}{G(t)} = \frac{e^{-(t+s)\lambda}}{e^{-\lambda t}} = e^{-\lambda s} = \Pr\left[X > s\right].$$

If X is interpreted as the lifetime of a cell, then the above equation tells: the (conditional) probability, given that the cell has lived for a length t of time, that it will survive a further length s of time, is given by the (unconditioned) probability of surviving duration s, thus forgetting that it has survived duration t. In other words an exponential random variable X has the *memoryless property*

$$\Pr\left[X > t + s \mid X > t\right] = \Pr\left[X > s\right]. \tag{4.9}$$

In fact, any random variable satisfying the memoryless property must be exponential. To show this, we write

$$G(t + s) = \Pr\left[X > t + s\right] = \Pr\left[X > t + s \cap X > t\right]$$
$$= \Pr\left[X > t\right] \Pr\left[X > t + s \mid X > t\right]$$
$$= \Pr\left[X > t\right] \Pr\left[X > s\right] = G(t)G(s),$$

where the last step follows from the memoryless property. For $t = 0$, $G(s) = G(0)G(s)$ implies $G(0) = 1$ if $G(s) > 0$, which is the nontrivial case. Since the derivatives of $G(t + s)$ with respect to t, s, and $t + s$ are all the same, we have

$$G'(t + s) = G'(t)G(s) = G(t)G'(s),$$

giving

$$\frac{G'(t)}{G(t)} = \frac{G'(s)}{G(s)}.$$

But this is possible only if both sides are equal to a constant, say $-\lambda$,

$$\frac{G'(t)}{G(t)} = -\lambda,$$

with the only nontrivial solution

$$G(t) = e^{-\lambda t},$$

which is bounded for $\lambda > 0$ and corresponds to the exponential random variable, Exponential (λ), and it follows that any nonnegative continuous random variable satisfying the memoryless property (4.9) has an exponential distribution. In other words, the exponential distribution is the only continuous distribution that satisfies the memoryless property. This result is very important and will be used in the section on Markov processes to find the distribution of the interevent times.

4.2.4 Mixed-Type Random Variables

For a random variable X of mixed type, taking values x from a continuous state space with density $p^{\mathrm{cont}}(x)$ and values n from a discrete state space with probability $P(n)$, the CDF $F(x)$ can be written as

$$F(x) = \int_{-\infty}^{x} p^{\mathrm{cont}}(x')\mathrm{d}x' + \sum_{n \leq x} P(n).$$

To find an expression for the density $p(x)$ of the random variable X of mixed type, differentiate both sides with respect to x,

$$p(x) = \frac{\mathrm{d}F(x)}{\mathrm{d}x} = p^{\mathrm{cont}}(x) + \sum_{n} P(n)\delta(x - n),$$

where the delta function $\delta(x)$ is defined by

$$\int_{-\infty}^{\infty} g(x)\delta(x - c)\mathrm{d}x = g(c)$$

for any (real) constant c and an arbitrary function g defined on the real axis. The probability that X falls in an event set B is then given by

$$\mathrm{Pr}(X \in B) = \int_{B} p(x)\mathrm{d}x = \int_{B} p^{\mathrm{cont}}(x)\mathrm{d}x + \sum_{n \in B} P(n).$$

This also shows that one can use probability density functions to characterize both continuous and discrete random variables.

4.3 Random Number Generation

Here follows a brief review of one way to compute a sample y of a random variable Y. The sample y is a random number selected from the probability distribution of Y. These concepts are important for computational experiments including the stochastic simulation algorithm (Chapter 5).

Bernoulli distribution: The standard uniform random variable U is very useful for generating random numbers of other distributions. To construct a Bernoulli variable Bernoulli (p) with a parameter $0 < p < 1$, we define

$$X = \begin{cases} 1 & \text{if } U < p, \\ 0 & \text{otherwise,} \end{cases}$$

so that, following (4.7),

$$\Pr[X = 1] = \Pr[U < p] = p,$$
$$\Pr[X = 0] = 1 - p.$$

This random variable X has a Bernoulli distribution with parameter p. If u is a uniform random number selected from the unit interval, then

$$x = \begin{cases} 1 & \text{if } u < p, \\ 0 & \text{otherwise,} \end{cases}$$

is a sample of Bernoulli (p).

Binomial distribution: Recall that a binomial variable Binomial (n, p) is the sum of n independent Bernoulli variables Bernoulli (p). To generate a Binomial random number, generate n independent Bernoulli variables with the common success probability p and then add them.

4.3.1 The CDF Inversion Method

The CDF
$$F(x) = \Pr[X \le x]$$
of a random variable X has values in the interval $[0, 1]$. This is also the range of the standard uniform random variable U. In an attempt to find a possible

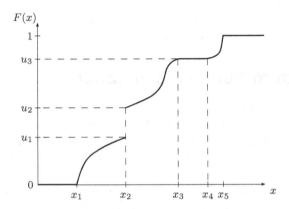

Figure 4.6 The CDF inversion method.

relationship between the two, we recall the unique property $\Pr[U \le u] = u$, which for $u = F(x)$ becomes

$$\Pr[U \le F(x)] = F(x).$$

If the CDF $F(x)$ is strictly increasing, its inverse $F^{-1}(u)$ at any point u is easily found to be the point on the abscissa corresponding to the point u on the ordinate. The CDF shown in Figure 4.6 is strictly increasing during the intervals (x_1, x_2), (x_2, x_3), and (x_4, x_5). For any u in one of the intervals $(0, u_1)$, (u_2, u_3), and $(u_3, 1)$, the inverse $F^{-1}(u)$ is simply the projection of u back to the abscissa. In such cases, the event $U \le F(x)$ is equivalent to $F^{-1}(U) \le x$, and we can write the above CDF as

$$\Pr\left[F^{-1}(U) \le x\right] = F(x).$$

But this is possible only if $X = F^{-1}(U)$. Thus any random variable with a known CDF $F(\cdot)$ can be expressed as the inverse function $F^{-1}(U)$ of the standard uniform random variable U. The requirement of $F(x)$ to be strictly increasing was necessary only for the above derivation and is not a serious issue because we can replace the inverse function by a quantile. The *u*th *quantile* or 100*u*th *percentile* of the distribution of X is the smallest number q_u such that

$$F(q_u) = \Pr[X \le q_u] = u.$$

The most familiar example is the *median* of a distribution, which is its half quantile or 50th percentile. Note that finding a quantile, unlike the inverse $F^{-1}(U)$, does not require the CDF to be strictly increasing. The CDF in

Figure 4.6 has a discontinuity at x_2 and a flat (nonincreasing) portion during (x_3, x_4). Although the CDF seems undefined at $x = x_2$, the probability $\Pr[X = x_2] = u_2 - u_1$ is well defined, and $F(x_2) = u_2$ by convention. Therefore, for every $u_1 \leq u \leq u_2$, the uth quantile is x_2. For the flat portion, u_3 maps to the interval $[x_3, x_4]$, from which we choose, by definition, the smallest number x_3 as the u_3th quantile. Returning to the problem of random number generation, if u is a uniform random number selected from the unit interval, then the uth quantile of the distribution of X is a sample of X. When the distribution function F has a closed-form expression, then finding the uth quantile is as easy as solving the equation $F(x) = u$ for x.

Exponential distribution: To generate a sample x of $X \sim \text{Exponential}(\lambda)$ from a standard uniform random number u, we need to solve

$$u = F(x) = 1 - e^{-\lambda x}$$

for x. The solution is

$$x = -\frac{1}{\lambda} \log(1 - u).$$

Since $1 - u$ is also a standard uniform random number, we could replace $1 - u$ by u, and the new number

$$\tau = -\frac{\log u}{\lambda} \tag{4.10}$$

is another sample of Exponential (λ).

4.3.2 Empirical Discrete Distribution

Consider a discrete random variable X taking values x with probabilities

$$P(x) = \Pr[X = x],$$

and the CDF

$$F(x) = \Pr[X \leq x] = \sum_{j \leq x} P(j).$$

Pick a sample u of the uniform random variable U in the interval $(0, 1]$ and set $u = F(x)$, which needs to be solved for x to give the desired random number. We observe that

$$\begin{aligned}
P(x) &= F(x) - F(x^-) \\
&= \Pr[U \leq F(x)] - \Pr[U \leq F(x^-)] \\
&= \Pr[F(x^-) < U \leq F(x)],
\end{aligned}$$

where $x^- = \max_m\{m < x\}$ is the state nearest to x from below. It follows that x is a sample of X if

$$F(x^-) < u \le F(x).$$

Thus computing x involves a search algorithm. Mathematically, the solution x can be written as

$$x = \inf_m\{m : F(m) \ge u\} = \inf_m\Big\{m : \sum_{j \le m} P(j) \ge u\Big\}. \qquad (4.11)$$

Matlab implementation: The seemingly complicated equation above can be implemented in a compact piece of Matlab code. To generate a sample of a discrete random variable X taking values 2, 3, and 4 with respective probabilities 0.5, 0.2, and 0.3, we could write the following code:

```
X = [2 3 4]; % state space
P = [0.5 0.2 0.3]; % probabilities
u = rand; % uniform random number
F = cumsum(P); % CDF
x = X(find(F>=u,1)); % generated random number
```

4.4 Random Vectors

One can study a vector

$$X = (X_1, \ldots, X_s)^T$$

of random variables defined on a common probability space. The vector-valued mapping

$$X : \Omega \to \mathbb{R}^s \qquad \text{with assignment} \qquad \omega \mapsto X(\omega)$$

is called a multivariate random variable or a *random vector*, which simply means that each component X_i is a random variable (discrete or continuous) taking real values x_i. For a given $\omega \in \Omega$, the quantity

$$x = X(\omega) = \Big(X_1(\omega), \ldots, X_s(\omega)\Big)^T$$

is a (sample) realization of the random vector. The joint cumulative distribution function of X can be defined as

$$F(x_1, \ldots, x_s) = \Pr\Big[\bigcap_{i=1}^{n} X_i \le x_i\Big],$$

which can be written in terms of the joint probability mass function $P(x_1, \ldots, x_s)$ as

$$\sum_{x_1' \leq x_1} \cdots \sum_{x_s' \leq x_s} P(x_1, \ldots, x_s)$$

when X is discrete, and in terms of the joint probability density $p(x_1, \ldots, x_s)$ as

$$\int_{-\infty}^{x_1} \cdots \int_{-\infty}^{x_s} p(x_1', \ldots, x_s') \, dx_1' \cdots dx_s'$$

when X is continuous. We introduce the following simplified notation:[2]

$$F(x) = F(x_1, \ldots, x_s),$$
$$P(x) = P(x_1, \cdots, x_s),$$
$$p(x) = p(x_1, \ldots, x_s),$$
$$\sum_{x' \leq x} = \sum_{x_1' \leq x_1} \cdots \sum_{x_s' \leq x_s},$$
$$\int_{x' \leq x} dx' = \int_{-\infty}^{x_1} dx_1' \cdots \int_{-\infty}^{x_s} dx_s'.$$

This notation allows us to write

$$F(x) = \begin{cases} \displaystyle\sum_{x' \leq x} P(x'), & \text{for discrete } X, \\[3mm] \displaystyle\int_{x' \leq x} p(x') dx', & \text{for continuous } X. \end{cases}$$

The probability for the random vector X to fall into a set $B \subset \mathbb{R}^s$ is then given by

$$\Pr(X \in B) = \begin{cases} \displaystyle\sum_{x \in B} P(x), & \text{for discrete } X, \\[3mm] \displaystyle\int_{x \in B} p(x) dx, & \text{for continuous } X. \end{cases}$$

---

[2]This notation can mislead the reader to conclude that $dx = (dx_1, \ldots, dx_s)$ from $x = (x_1, \ldots, x_s)$, but here dx denotes the product $dx_1 \cdots dx_s$.

4.5 Expectations

The probability distribution of a random variable X fully specifies its behavior. However, one is often interested in the average behavior of X or some function $g(X)$. This motivates the following definition. If X is a random variable taking real values x with probability distribution $F(x)$, which corresponds to a probability mass $P(x)$ or density $p(x)$ depending on whether X is discrete or continuous, then the *expectation*, or *mean*, of X is defined by

$$\langle X \rangle = \int x \, dF(x) = \begin{cases} \displaystyle\sum_n n P(n), & \text{for discrete } X, \\[2ex] \displaystyle\int x \, p(x) dx, & \text{for continuous } X, \end{cases}$$

where the integration is understood to be over all possible values $x \in X(\Omega)$, the summation is over all values from the discrete state space, and the last integration is over all values from the continuous state space. Note that the expectation is a linear operator, because both the summation and integration are linear operators. More generally, the expectation of some measurable function $g(X)$ of X is defined to be

$$\langle g(X) \rangle = \int g(x) \, dF(x) \, .$$

Particularly important expectations are the moments of order m:

$$\langle X^m \rangle = \int x^m \, dF(x) \, .$$

The deviation of a random variable X from its mean value will be written as

$$\delta X = X - \langle X \rangle \, .$$

The *variance* of X is then defined by

$$\langle \delta X^2 \rangle \stackrel{\text{def}}{=} \left\langle (\delta X)^2 \right\rangle = \left\langle (X - \langle X \rangle)^2 \right\rangle \, .$$

The significance of the variance stems from its property as a measure of the fluctuations of the random variable X, that is, the extent of deviations of the realizations of X from the mean value. The linearity of the expectation operator leads to the useful relation

$$\langle \delta X^2 \rangle = \langle X^2 \rangle - \langle X \rangle^2 \, . \tag{4.12}$$

An alternative measure of fluctuation is the standard deviation, which is the square root of the variance,

$$\text{SD}(X) = \sqrt{\langle \delta X^2 \rangle},$$

which is notable for its usefulness as a measure of fluctuation in units of X, the coefficient of variation.

Poisson moments: If $X \sim \text{Poisson}(\lambda)$, with distribution (4.6), then for the mean,

$$\langle X \rangle = \sum_{x=0}^{\infty} x \frac{e^{-\lambda}\lambda^x}{x!} = \lambda \sum_{x=1}^{\infty} \frac{e^{-\lambda}\lambda^{x-1}}{(x-1)!} = \lambda \sum_{y=0}^{\infty} \frac{e^{-\lambda}\lambda^y}{y!} = \lambda,$$

where the last step uses the normalization of probability to unity. For the variance, we need the second moment, for which the following expectation is helpful:

$$\langle (X-1)X \rangle = \sum_{x} x(x-1)p(x)$$

$$= \sum_{x=1}^{\infty} x(x-1) \frac{e^{-\lambda}\lambda^x}{x!}$$

$$= \lambda^2 \sum_{x=2}^{\infty} \frac{e^{-\lambda}\lambda^{x-2}}{(x-2)!}$$

$$= \lambda^2.$$

Since expectation is a linear operator, we have $\langle (X-1)X \rangle = \langle X^2 \rangle - \langle X \rangle$, from which follows

$$\langle X^2 \rangle = \langle (X-1)X \rangle + \langle X \rangle = \lambda^2 + \lambda.$$

The variance can now be determined:

$$\langle \delta X^2 \rangle = \langle X^2 \rangle - \langle X \rangle^2 = \lambda^2 + \lambda - \lambda^2 = \lambda.$$

Thus the parameter λ gives both the mean and variance of the Poisson random variable.

Exponential moments: If $X \sim$ exponential(λ), then X has the mean

$$\langle X \rangle = \int_0^\infty x\lambda e^{-\lambda x}\,\mathrm{d}x = \int_0^\infty e^{-\lambda x}\,\mathrm{d}x = \frac{1}{\lambda},$$

which is why λ is called the rate of X. Moreover, X has second moment

$$\langle X^2 \rangle = \int_0^\infty x^2\lambda e^{-\lambda x}\,\mathrm{d}x = \frac{2}{\lambda}\int_0^\infty xe^{-\lambda x}\,\mathrm{d}x = \frac{2}{\lambda^2}$$

and variance

$$\langle \delta X^2 \rangle = \langle X^2 \rangle - X^2 = \frac{2}{\lambda^2} - \left(\frac{1}{\lambda}\right)^2 = \frac{1}{\lambda^2}.$$

4.5.1 Multidimensional Expectations

Consider a random vector $X = (X_1, \ldots, X_n)^T$ with state space $X(\Omega) \subset \mathbb{R}^n$ and distribution function $F(x)$ that corresponds to a joint probability density $p(x)$ or a mass $P(x)$ depending on whether X is continuous or discrete. The componentwise expectation

$$\langle X_i \rangle = \int x_i\,\mathrm{d}F(x)$$

can be written as

$$\langle X_i \rangle = \sum_x x_i P(x)$$

when X is discrete, and

$$\langle X_i \rangle = \int x_i p(x)\mathrm{d}x$$

when X is continuous. In terms of the moments

$$\langle X_i X_j \rangle = \int x_i x_j\,\mathrm{d}F(x),$$

the *covariance matrix* $\langle \delta X \delta X^T \rangle$ is defined elementwise by covariances

$$\langle \delta X_i \delta X_j \rangle = \langle (X_i - \langle X_i \rangle)(X_j - \langle X_j \rangle) \rangle = \langle X_i X_j \rangle - \langle X_i \rangle \langle X_j \rangle.$$

The covariance is zero when X_i and X_j are independent, because then $\langle X_i X_j \rangle = \langle X_i \rangle \langle X_j \rangle$, which follows from independence. In matrix notation, we can write

$$\langle \delta X \delta X^T \rangle = \left\langle (X - \langle X \rangle)(X - \langle X \rangle)^T \right\rangle = \langle XX^T \rangle - \langle X \rangle \langle X \rangle^T . \quad (4.13)$$

Here the superscript T denotes transpose of a matrix. In the next section we introduce generating functions, a further important class of expectations that may serve to characterize completely a random variable.

4.5.2 Generating Functions

Consider a random variable X taking values x from some state space with probability density $p(x)$ when continuous or mass $P(x)$ when discrete. We introduce the expectation, for complex z,

$$\mathcal{P}(z) = \langle z^X \rangle,$$

which is a sum

$$\mathcal{P}(z) = \sum_k z^k P(k)$$

when X is discrete and an integral

$$\mathcal{P}(z) = \int z^x p(x) \mathrm{d}x$$

when X is continuous. The function $\mathcal{P}(z)$ is well defined for $|z| \leq 1$, since both the above sum and the integral converge in that region. If X takes only nonnegative integer values k, then the above sum is a Taylor expansion,

$$\mathcal{P}(z) = \sum_{k=0}^{\infty} z^k P(k).$$

Note how probabilities $P(k)$ can be recovered from $\mathcal{P}(z)$ as coefficients of the above expansion, which is why $\mathcal{P}(z)$ is referred to as the *probability generating function* (PGF) of the distribution. The Taylor coefficients satisfy

$$P(k) = \frac{1}{k!} \mathcal{P}^{(k)}(0) = \frac{1}{2\pi \mathrm{i}} \oint_{C(0)} \frac{\mathcal{P}(z)}{z^{k+1}} \mathrm{d}z,$$

where $\mathcal{P}^{(k)}(0)$ is the nth derivative of $\mathcal{P}(z)$ evaluated at $z = 0$ and $C(0)$ denotes a contour around $z = 0$ and inside the unit circle. The PGF can also be used to calculate the mean and variance of X. The reader is encouraged to show that

$$\langle X \rangle = \mathcal{P}'(1)$$

for the mean and

$$\langle \delta X^2 \rangle = \mathcal{P}''(1) + \mathcal{P}'(1) - [\mathcal{P}'(1)]^2$$

for the variance (provided the mean is finite). The PGF is not useful if X is not restricted to nonnegative integers. In that case it is useful to use $z = e^w$ to have the expectation

$$M(w) = \mathcal{P}(e^w) = \langle e^{wX} \rangle .$$

The mth derivative of $M(w)$,

$$M^{(m)}(w) = \langle X^m e^{wX} \rangle ,$$

gives the mth-order moment when $w = 0$,

$$M^{(m)}(0) = \langle X^m \rangle ,$$

which shows that $M(w)$ is the *moment generating function* (MGF) for the random variable X. Sometimes it is more convenient to use $w = iu$, where $i = \sqrt{-1}$ and the resulting function $\phi(u) = \langle e^{iuX} \rangle$, called the *characteristic function,* is well defined for all real values of u. If X is continuous, then the MGF

$$M(w) = \int_{-\infty}^{\infty} e^{wx} p(x) dx \tag{4.14}$$

can be used to generate the density function through the inverse transform,

$$f(x) = \frac{1}{2\pi i} \int_{c-i\infty}^{c+i\infty} e^{-wx} M(w) dw,$$

where c is a real number such that the integral in (4.14) converges to the left of $w = c$. If X takes only takes nonnegative values, the integral in (4.14) converges in the left half-plane, and we can choose $c = 0$.

Another useful expectation is the natural logarithm of the MGF,

$$K(w) = \log M(w) .$$

The first two derivatives of $K(w)$ directly give the mean and variance:

$$K'(0) = \langle X \rangle , \quad K''(0) = \delta X .$$

Higher derivatives of $K(w)$ are combinations of the moments. The mth derivative of $K(w)$ evaluated at the origin is called the mth *cumulant* of the random variable X,

$$\kappa_m = K^m(0),$$

and the function $K(w)$ itself is called the *cumulant generating function* (CGF) for obvious reasons. The natural logarithm of the characteristic function behaves exactly the same. In terms of the generating functions, the mean of a discrete random variable can be expressed as

$$\langle X \rangle = \mathcal{P}'(1) = M'(0) = K'(0),$$

and the variance as

$$\langle \delta X^2 \rangle = \mathcal{P}''(1) + \mathcal{P}'(1) - [\mathcal{P}'(1)]^2 = M''(0) - [M'(0)]^2 = K''(0)$$

Poisson generating functions: If $X \sim \text{Poisson}(\lambda)$, then X has PGF

$$\mathcal{P}(z) = \sum_{n=0}^{\infty} z^n \frac{e^{-\lambda}\lambda^n}{n!} = e^{-\lambda} \sum_{n=0}^{\infty} \frac{n}{n!} = e^{-\lambda}e^{\lambda z} = e^{(z-1)\lambda},$$

MGF

$$M(w) = \mathcal{P}(e^w) = \exp\left((e^w - 1)\lambda\right),$$

and CGF

$$K(w) = \log M(w) = (e^w - 1)\lambda.$$

The mean and variance of X can be computed from these generating functions:

$$\langle X \rangle = K'(0) = \lambda, \qquad \langle \delta X^2 \rangle = K''(0) = \lambda.$$

Exponential MGF and random sums: If $X \sim \text{exponential}(\lambda)$, then X has the MGF

$$M(w) = \int_0^{\infty} e^{wx}\lambda e^{-\lambda x}\,dx = \lambda \int_0^{\infty} e^{(w-\lambda)x}\,dx = \frac{\lambda}{\lambda - w},$$

which converges for $|w| < \lambda$.

If we have a random vector $X = (X_1, \ldots, X_n)^T$, with probability density $p(x)$ when continuous or mass $P(x)$ when discrete, the definitions of generating functions can be extended as

$$\mathcal{P}_X(z) = \mathcal{P}_{X_1 \dots X_n}(z_1, \dots, z_n) = \left\langle \prod_{i=1}^{n} z_i^{X_i} \right\rangle,$$

$$M_X(w) = M_{X_1 \dots X_n}(w_1, \dots, w_n) = \left\langle \exp \left(\sum_{i=1}^{n} w_i X_i \right) \right\rangle,$$

both involving integrals (or sums) of the argument of $\langle \cdot \rangle$, weighted by $p(x)$ (or p_x), over all possible values of X. If X_i are independent, then we have

$$\left\langle \prod_{i=1}^{n} z_i^{X_i} \right\rangle = \prod_{i=1}^{n} \left\langle z_i^{X_i} \right\rangle, \quad \left\langle \exp \left(\sum_{i=1}^{n} w_i X_i \right) \right\rangle = \prod_{i=1}^{n} \left\langle e^{w_i X_i} \right\rangle,$$

and therefore

$$\mathcal{P}_X(z) = \prod_{i=1}^{n} \mathcal{P}_{X_i}(z_i) \quad \text{and} \quad M_X(w) = \prod_{i=1}^{n} M_{X_i}(w_i).$$

Setting $z_i = z$, $w_i = w$ for all i, and $Y = \sum_{i=1}^{n} X_i$, we get

$$\mathcal{P}_Y(z) = \langle z^Y \rangle = \prod_{i=1}^{n} \mathcal{P}_{X_i}(z) \quad \text{and} \quad M_Y(w) = \langle e^{wY} \rangle = \prod_{i=1}^{n} M_{X_i}(w). \quad (4.15)$$

Poisson random sums: If $(X_1, \dots, X_n)$ are independent Poisson distributed random variables with rate parameters $\lambda_1, \dots, \lambda_n$, and and $Y = \sum_{i=1}^{n} X_i$, then (4.15) combined with $\mathcal{P}_{X_i}(z) = e^{(z-1)\lambda_i}$ gives

$$\mathcal{P}_Y(z) = \prod_{i=1}^{n} \mathcal{P}_{X_i}(z) = \prod_{i=1}^{n} e^{(z-1)\lambda_i} = \exp \left((z-1) \sum_{i=1}^{n} \lambda_i \right),$$

which shows that

$$Y = \sum_{i=1}^{n} X_i \sim \text{Poisson} \left(\sum_{i=1}^{n} \lambda_i \right). \quad (4.16)$$

Thus the sum of independent Poisson random variables is also Poisson distributed with rate parameter that is the sum of the individual rate parameters.

Exponential random sums: If $X_1, \dots, X_n$ are independent each being Exponential (λ), and $Y = \sum_{i=1}^{n} X_i$, then (4.15) combined with $M_{X_i}(w) =$

$\lambda/(\lambda - w)$ gives

$$M_Y(w) = \prod_{i=1}^{n} M_{X_i}(w) = \left(\frac{\lambda}{\lambda - w}\right)^n.$$

The probability density of Y can be obtained from the inverse transform,

$$p_Y(y) = \frac{1}{2\pi i} \int_{c-i\infty}^{c+i\infty} e^{-wy} M_Y(w)\, dw = \frac{\lambda^n}{(-1)^n 2\pi i} \int_{c-i\infty}^{c+i\infty} \frac{e^{-wy}}{(w-\lambda)^n}\, dw,$$

which is a contour integral with nth order pole at $w = \lambda$. To perform this integration, we can use the Cauchy relation

$$\frac{1}{k!} G^{(k)}(\alpha) = \frac{1}{2\pi i} \oint_{C(\alpha)} \frac{G(w)}{k+1}\, dw.$$

In our case, $G(w) = e^{-wy}$, $k = n - 1$, and $\alpha = \lambda$, from which $G^{(k)}(\alpha) = G^{(n-1)}(\lambda) = (-1)^n y^{n-1} e^{-\lambda y}$. Combining all the above leads to the PDF

$$p_Y(y) = \frac{\lambda^n y^{n-1}}{(n-1)!} e^{-\lambda y}$$

of the sum Y of n independent variables each exponentially distributed with a common rate λ. This is the Erlang distribution defined in (3.9) with rate parameter λ and shape parameter n. We write $Y \sim \text{Erlang}(\lambda, n)$ to denote that Y follows the Erlang distribution. We can summarize this important result as follows

$$\text{if } X_i \sim \text{Exponential}(\lambda), \quad \text{then} \quad \sum_{i=1}^{n} X_i \sim \text{Erlang}(\lambda, n). \tag{4.17}$$

This result has a very interesting implication: If X_i is interpreted as the time between two consecutive events from a sequence of events, the occurrences of which, over nonoverlapping intervals, are independent, then $Y = \sum_{i=1}^{n} X_i$ represents the waiting time until the occurrence of the nth event, starting from zero. Thus, the waiting time until the nth event follows an Erlang distribution with rate parameter λ and shape parameter n.

The gamma distribution: The factorial in the Erlang density is a special case of the more general gamma function, in light of the well-known relation

$$\Gamma(n) = (n-1)!.$$

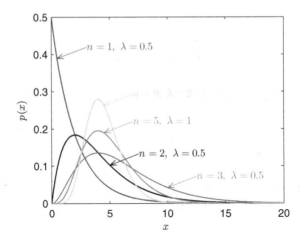

Figure 4.7 Gamma distribution for five different value pairs of the rate parameter λ and the shape parameter n. Note how a range of simpler distributions can be represented by altering the values of the two parameters.

Since the gamma function is defined for every positive real number n, we can generalize the Erlang distribution to the *gamma distribution* denoted by Gamma (λ, n), which has probability density

$$p_Y(y) = \frac{\lambda^n y^{n-1}}{\Gamma(n)} e^{-\lambda y} . \tag{4.18}$$

Figure 4.7 illustrates how the gamma distribution can represent a range of simpler distributions for different values of the rate parameter λ and the shape parameter n.

4.6 Stochastic Processes

A *stochastic process* X is a map that assigns a real number (or an n-vector) to each outcome ω of the sample space Ω for each element t of an index set I,

$$X : \Omega \times I \to \mathbb{R}^n, \qquad (\omega, t) \mapsto X(\omega, t), \quad t \in I, \, \omega \in \Omega .$$

The process is univariate for $n = 1$ and multivariate for $n > 1$. In most physical applications, the parameter t plays the role of the *time* variable, in which case I is the nonnegative real axis. Each value $X(\omega, t)$ is the *state* of the process, and the set of all possible values forms the *state space* of the process. The interpretation of a stochastic process X depends on the variation of ω and t. Thus X is interpreted as

1. A *family* (or an *ensemble*) of functions $\{X(\omega, t) : \omega \in \Omega, t \in I\}$ when both t and ω are variables.

2. A *realization, trajectory,* or *sample path* $X_\omega(t)$ of the stochastic process when t is a variable and ω is fixed.

3. A *random variable (or vector)* $X_t(\omega)$ of the process at time t when ω is variable and t is fixed.

4. Simply a number (or vector) when both t and ω are fixed.

It is common practice to write $X(t)$ for a stochastic process, and the dependence on ω is taken for granted. If the index set I is a countable set, $X(t)$ is a discrete-time process. If I is a continuum (e.g., real line), $X(t)$ is a continuous-time process. If the state space is a countable set, $X(t)$ is a discrete-state process. A discrete-state stochastic process is also called a *chain*. If the state space is a continuum, $X(t)$ is a continuous-state process. For example, the outcome of n tosses of a coin is a discrete-time chain with state space $\{\text{head}, \text{tail}\}$ and time index $I = \{0, 1, 2, \ldots, n\}$. The number of telephone calls in a region during a time interval $[a, b]$ and the number of transcription initiation cycles in a promoter region during a time interval $[a, b]$ are examples of a continuous-time chain because $t \in [a, b]$. The daily measurement of temperature or hourly measurement of concentration of a protein in a cell are examples of a discrete-time continuous-state process. The position of a Brownian particle (small pollen grain suspended in water) at a certain time t is a continuous-time continuous-state (stochastic) process. Note that a stochastic process is a mathematical abstraction and should not be confused with the physical process being modeled.

In continuous-time stochastic processes, it is convenient to define *increments* as the differences $X(t) - X(s)$, which are random variables for $s < t$. The continuous-time stochastic process $X(t)$ is said to have *independent increments* if changes $X(t_1) - X(t_0)$, $X(t_2) - X(t_1)$, $\ldots$ in the state of the process in different time intervals $t_0 < t_1 < \cdots$ are independent. If changes $X(t + w) - X(t)$ in the value of the process depend only on the lengths w during which the change occurs, not on the time points t of measuring the change, then the continuous-time stochastic process is said to have *stationary increments*. Stochastic processes are distinguished by three principal properties:

1. the state space $X(\Omega, I)$,

2. the index set I,

3. the dependence relations between the random variables $X(t)$ at different times.

An alternative name for a stochastic process is *random process*. The level of randomness in a process is determined by the dependence relations between the random variables. Thus a process with independent and identically distributed random variables is as random as it can be. For example, a Poisson process, defined later in this sequel, is a continuous-time (stochastic) chain $X(t)$ taking nonnegative integer values at times $t \geq 0$, and has stationary and independent increments $X(t + s) - X(t)$ with a Poisson distribution.

A generalization of the Poisson process is a *counting process*, defined as a (stochastic) chain $X(t)$ taking nonnegative integer values; it represents the total number of events that have occurred in a time interval $[0, t]$. Examples of counting processes are, the number of telephone calls at a local exchange, the number of transcription initiation cycles in a promoter region, and the number of virus attacks on a computer, all during a time interval. Processes with independent and identically distributed random variables are not always interesting as stochastic models because they behave more or less in the same way. However, stochastic processes that allow some dependence on the past can give variability in their behavior, and that is why the dependence relationship is so important. The dependence relations between the random variables of a particular stochastic process are usually derived from the modeling assumptions made about the process for analysis. Typically the process is specified by determining its local behavior (in a short time interval next to the current time), and the goal of our analysis is to discover its global behavior. In most cases, it is reasonable to assume that the local behavior depends only on the current state of the process and not on the past. This property is called the *memoryless* or the *Markov property,* and a stochastic process having this property is a *Markov process.*

Construction: In practice, a stochastic process is constructed from experimental data. Take a set $t_1, \ldots, t_m$ of discrete times and Borel sets $B_1, \ldots, B_m$ in $\mathbb{R}^n$, and consider for a multivariate stochastic process $X(t)$ the joint probability distribution *of order m*,

$$\Pr\left[\bigcap_{i=1}^{m} X(t_i) \in B_i\right].$$

The process can then be specified by collecting these distributions for $m = 1, 2, \ldots$ to form the *family of finite joint probability distributions* of the stochastic process. A sample in Figure 4.8 illustrates the idea. The Borel sets B_i could be suitable intervals (bins) in a frequency histogram constructed from data.

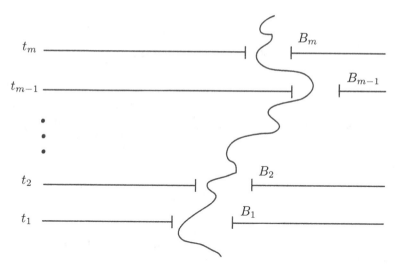

Figure 4.8 A sample path of a stochastic process. Figure adopted from [18].

In the next section, we introduce one of the most important stochastic processes, the Poisson process.

4.6.1 Poisson Process

Consider a sequence of events occurring at random times. Typical examples are

- Molecules arriving at (and passing through) a biological channel such as a pore in a cell membrane, as depicted in Figure 4.9,

- Transcription factors (arriving and) binding to a promoter region to initiate transcription,

- Students arriving in a classroom/library,

- Failure of a device (and being replaced by a new one).

Such a sequence of *arrivals* or *jumps* can be modeled by a continuous-time discrete-state stochastic process. Counting the number of such arrivals in a given time interval leads us to define a counting process. A counting process $X(t)$ represents the total number of arrivals in a time interval $[0, t]$, where $t \geq 0$. In every sample path of the counting process, the counter $X(t)$ starts in state $X(0) = 0$ and is incremented by size one on each arrival. Thus the state space of a counting process is the set of nonnegative integers. The nth *arrival time* is denoted by W_n, and since $X(0) = 0$ is not an arrival, $W_0 = 0$

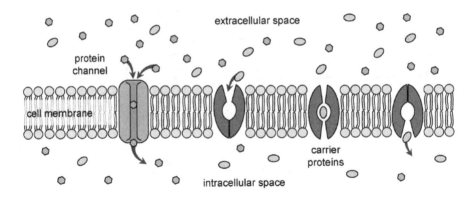

Figure 4.9 Molecules arriving at a channel in the cell membrane. The channels in the cell membrane provide a mechanism to selectively pass molecules arriving from the extracellular space into the intracellular space.

by definition. Alternative names used in the literature for the arrival time are "jump time" and "waiting time." The *interarrival time* between the nth and the next arrival is $T_n = W_{n+1} - W_n$. Alternative names used in the literature for the interarrival time are "holding time" and "sojourn time." Figure 4.10 shows a sample path of a counting process, illustrating the arrival times and the interarrival times. Note that since the nth jump is, by definition, to state n, the two events $W_n \leq t$ and $X(t) \geq n$ are identical. Therefore a counting process can be defined in terms of the arrival times:

$$X(t) = \max_n \{n : W_n \leq t\}, \qquad \text{where} \quad W_n = \sum_{i=1}^{n-1} T_i. \qquad (4.19)$$

In some cases, for example when a device fails and is replaced by a new one, it may be reasonable to assume that the process probabilistically restarts (or renews) at each arrival. This *renewal property* means that the interarrival times T_k are independent and identically distributed random variables. A counting process that has independent and identically distributed interarrival times is thus called a *renewal process*. Since the common distribution of the interarrival times can be arbitrary, a general renewal process has a *weak* renewal property. The interested reader can find further information in [2, 5].

Note that a renewal process is more random than a general counting process. We can get even more randomness if the process has the memoryless property, which is the case if the arrivals are not coordinated with each other. A renewal process with the memoryless property is called a *Poisson process*.

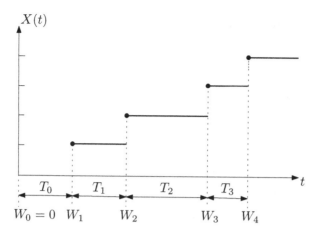

Figure 4.10 Illustration of events taking place in a counting process. At each arrival at time $t = W_n$, the quantity $X(t)$ is incremented by unity (hence the qualification counting process). The interarrival times T_n are continuous random variables. They are independent and identically distributed for a renewal process, and, further, are exponentially distributed for a Poisson process. For the latter, the arrival times W_n have the gamma distribution because each arrival time is a sum of the exponential interarrival times T_n. Recall the property (4.17) of independent and identically distributed exponential random variables. Figure adopted from [5].

Thus a Poisson process is a counting process $X(t)$ that has independent and identically distributed exponentially distributed interarrival times. Since the common distribution of interarrival times is now fixed, the Poisson process has the *strong* renewal property. If λ is the rate parameter of the common exponential distribution, we can write, for the interarrival times of a Poisson process,

$$T_n \sim \text{Exponential}\,(\lambda), \quad n = 0, 1, 2, \dots .$$

It follows from (4.17) that

$$W_n = \sum_{i=0}^{n-1} T_i \sim \text{Erlang}\,(\lambda, n)$$

for the time of the nth arrival. The distribution function of W_n can be obtained by integrating its density (4.18),

$$\Pr\,[W_n \leq t] = \int_0^t \frac{\lambda^n y^{n-1}}{(n-1)!} e^{-\lambda y} dy\,.$$

Integrating by parts, we can obtain the following recurrence relation:

$$\Pr[W_n \leq t] = \Pr[W_{n+1} \leq t] + \frac{(\lambda t)^n e^{-\lambda t}}{n!}.$$

Hence we can write for $n > 0$,

$$\begin{aligned}
\Pr[X(t) = n] &= \Pr[X(t) \geq n] - \Pr[X(t) \geq n+1] \\
&= \Pr[W_n \leq t] - \Pr[W_{n+1} \leq t] \\
&= \frac{(\lambda t)^n e^{-\lambda t}}{n!},
\end{aligned}$$

and for $n = 0$,

$$\Pr[X(t) = 0] = \Pr[T_0 > t] = e^{-\lambda t},$$

which shows that $X(t) \sim \text{Poisson}(\lambda t)$ and why the process is called a Poisson process. Next we investigate how the increments $X(s+t) - X(s)$ are distributed. Given that the process is in state i at time $s > 0$, the time T^s until the next arrival, is also Exponential(λ). To see this, note that $X(s) = i$ is the same as $T_i > s - W_i$ and $T^s > t$ is the same as $T_i > s - W_i + t$. Therefore from the memoryless property of interarrival times, we can write

$$\begin{aligned}
\Pr[T^s > t \mid X(s) = i] &= \Pr[T_i > s - W_i + t \mid T_i > s - W_i] \\
&= \Pr[T_i > t] \\
&= e^{-\lambda t},
\end{aligned}$$

which establishes the result. Note that this result does not rely on the choice of i and s and hence is true in general,

$$\Pr[T^s > t] = e^{-\lambda t}.$$

It follows that $T^s, T_{i+1}, T_{i+2}, \ldots$ is a sequence of independent and identically distributed exponential interarrival times, in terms of which the increment $X(s+t) - X(s)$ can be expressed, for $t > 0$, as

$$X(s+t) - X(s) = \max_n \left\{ n : T^s + \sum_{r=1}^{n-1} T_{i+r} \leq t \right\},$$

which can be interpreted as a counting process with arrival times

$$W_n = T^s + \sum_{r=1}^{n-1} T_{i+r}, \quad n = 1, 2, 3, \ldots.$$

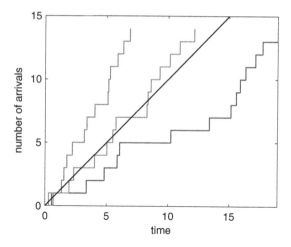

Figure 4.11 Three sample paths of the Poisson process with rate parameter $\lambda = 1$, together with the time-dependent mean λt (the straight line) of the process. Figure adopted from [5].

When compared with (4.19), these arrival times differ from those of the counting process $X(t)$ in the starting time T^s. Hence the process defined by the Poisson increments $X(s+t) - X(s)$ shares the same distribution with the counting process $X(t)$, that is,

$$X(s+t) - X(s) \sim X(t).$$

The Poisson increments $X(s+t) - X(s)$ are *stationary* because they do not depend on the choice of s, and are *independent* because the above property does not depend on the arrival times outside the interval $[s, s+t]$. Thus a Poisson process can also be defined as a counting process that has stationary and independent increments. In summary, for a Poisson process, the probability of n arrivals in a length t of time is given by

$$\Pr\left[X(t) = n\right] = \frac{(\lambda t)^n e^{-\lambda t}}{n!}, \quad n = 0, 1, 2, \ldots, \tag{4.20}$$

which corresponds to the Poisson distribution with rate parameter λt. Sample paths of a Poisson process can be computed by generating a sequence of exponentially distributed random numbers using (4.10) and incrementing the process by unity each time. Three sample paths of a Poisson process, with rate parameter $\lambda = 1$, are shown in Figure 4.11 together with the time-dependent mean λt of the process (plotted by a solid line).

Sample paths of the Poisson process can be easily computed in Matlab. The three sample paths in Figure 4.11 were generated by the following Matlab code:

```
% T = exprnd(1,14,1);
% if statistics toolbox is installed, uncomment the
% previous line and comment out the following line.
T = repmat(-1,14,1).*log(rand(14,1));
W = [0; cumsum(T)];
n = (0:14)';
stairs(W,n)
```

Our view about the Poisson process so far has been global. For general stochastic processes, this is not convenient, and the usual way is to look (locally) at what happens during an arbitrarily short time interval $(t, t + h]$. Since the Poisson increments are stationary, we can choose the interval to be $(0, h]$. We will use the notation $o(h)$ to represent negligible quantities that vanish faster than h as the latter approaches zero. More precisely, a function $f(h)$ is said to be $o(h)$ if $f(h)/h \to 0$ as $h \to 0$. The probability of no arrivals in the interval is

$$\Pr[X(h) = 0] = \Pr[T_0 > h] = e^{-\lambda h} = 1 - \lambda h + o(h),$$

and the probability of one arrival is

$$\Pr[X(h) = 1] = \Pr[T_0 \leq h] = 1 - e^{-\lambda h} = \lambda h + o(h).$$

The probability of two arrivals is

$$\Pr[X(h) = 2] = \Pr[T_0 + T_1 \leq h].$$

But the event $T_0 + T_1 \leq h$ is the same as the intersection of the two independent events $T_0 \leq \alpha h$ and $T_1 \leq (1 - \alpha)h$ for any real $0 \leq \alpha \leq 1$. Therefore

$$\begin{aligned}
\Pr[X(h) = 2] &= \Pr[T_0 \leq \alpha h] \Pr[T_1 \leq (1 - \alpha)h] \\
&= (\lambda \alpha h + o(h))((1 - \alpha)\lambda h + o(h)) \qquad (4.21) \\
&= o(h),
\end{aligned}$$

which shows that multiple arrivals in a short time interval are extremely rare. Let $P(n, t)$ denote the probability of n arrivals in a time interval of length t, that is,

$$P(n, t) = \Pr[X(t) = n].$$

Then the local behavior of a Poisson process in a short interval of length h can be summarized as

$$P(n, h) = o(h) + \begin{cases} 1 - \lambda h, & \text{if } n = 0, \\ \lambda h, & \text{if } n = 1, \\ 0, & \text{if } n > 1, \end{cases} \tag{4.22}$$

which tells that the arrivals in a Poisson process are independent over nonoverlapping intervals, have a constant density (or rate) λ, and that multiple arrivals in a short time interval are extremely rare. Let us investigate, for arbitrarily small h, the probability $P(n, t + h)$ of k arrivals in time $t + h$. For $k = 0$ it is straightforward. The probability of no arrivals in time $t + h$ is

$$P(0, t + h) = P(0, t)P(0, h)$$

because of the independence of arrivals in the nonoverlapping intervals $(0, t]$ and $(t, t + h]$. Using (4.22), we get

$$P(0, t + h) = P(0, t)(1 - \lambda h + o(h)),$$

and for vanishingly small h, we get the differential equation

$$\frac{\partial}{\partial t} P(0, t) = -\lambda P(0, t),$$

subject to the initial condition $P(0, 0) = \Pr[X(0) = 0] = 1$. The solution $P(0, t) = e^{-\lambda t}$ satisfies the Poisson distribution. For $k \geq 1$ we note that n arrivals in $t + h$ are possible in the following mutually exclusive ways:

1. There are n arrivals during $(0, t]$ and no arrivals during $(t, t + h]$. The probability of this happening is the product $P(n, t)P(0, h)$.

2. There are $n - 1$ arrivals during $(0, t]$ and a single arrival during $(t, t + h]$. The probability of this happening is the product $P(n - 1, t)P(1, h)$.

3. There are fewer than $n - 1$ arrivals during $(0, t]$ and more than one arrival during $(t, t + h]$. The probability of this happening is very small and represent by $o(h)$.

Thus the probability of n arrivals during $t + h$ is the sum of the above probabilities,

$$P(n, t + h) = (1 - \lambda h + o(h)) P(n, t) + (\lambda h + o(h)) P(n - 1, t) + o(h),$$

and for vanishingly small h, we obtain the system of differential equations

$$\frac{\partial}{\partial t} P(n,t) = -\lambda P(n,t) + \lambda P(n-1,t), \qquad k = 1,2,\dots .$$

This can be combined with the case $n = 0$ to give

$$\frac{\partial}{\partial t} P(n,t) = \begin{cases} -\lambda P(0,t), & \text{if } n = 0, \\ -\lambda P(n,t) + \lambda P(n-1,t) & \text{if } n \geq 1. \end{cases} \tag{4.23}$$

We can solve this system by employing the probability generating function presented earlier in Section 4.5.2. Multiply by z^n and sum over n,

$$\sum_{n=0}^{\infty} z^n \frac{\partial}{\partial t} P(n,t) = -\lambda \sum_{n=0}^{\infty} z^n P(n,t) + \lambda \sum_{n=1}^{\infty} z^n P(n-1,t),$$

where we note that $P(n,t)$ is zero for negative values of n. Interchanging summation and differentiation, and recognizing the probability generating function

$$\mathcal{P}(z,t) = \sum_{n=0}^{\infty} z^n P(n,t),$$

leads to the partial differential equation

$$\frac{\partial}{\partial t} \mathcal{P}(z,t) = -\lambda \mathcal{P}(z,t) + z\lambda \mathcal{P}(z,t) = (z-1)\lambda \mathcal{P}(z,t).$$

This partial differential equation can be solved for $\mathcal{P}(z,t)$ subject to the initial condition

$$\mathcal{P}(z,0) = \sum_{n=0}^{\infty} z^n P(n,0) = P(0,0) = 1$$

to give

$$\mathcal{P}(z,t) = e^{(z-1)\lambda t} = e^{-\lambda t} \sum_{n=0}^{\infty} \frac{(z\lambda t)^n}{n!} .$$

Comparing this with the definition of PGF gives the PMF for a Poisson process,

$$P(n,t) = e^{-\lambda t} \frac{(\lambda t)^n}{n!}, \qquad n = 0,1,2,\dots,$$

and thus we have arrived at (4.20). A more detailed explanation of the method of generating functions for solving differential equations involving probabilities can be found in [2, 5]. Therefore, combined with independence of increments, (4.22) can be used as an alternative definition of a Poisson process. To get a

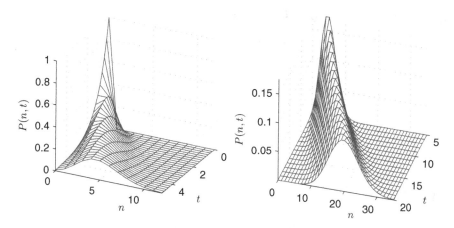

Figure 4.12 Temporal progress of the (distribution of a) Poisson process. The PMF is plotted for $\lambda = 1$ for two time intervals: $0 < t < 5$ (*left*) and $5 < t < 20$ (*right*).

feel for how a Poisson process evolves in time, the PMF of the process, with $\lambda = 1$, is plotted in Figure 4.12 at two time subintervals. The PMF spreads out from the initial peak and ultimately will look like a Gaussian distribution.

4.6.2 Birth–Death Processes

The arrivals of a Poisson process are so random that the current state n has no influence on the arrival rate λ. However, this is not an interesting behavior. For example, the number of individual cells in a culture does have an influence on the rate of proliferation or cell death. Such an influence could, for example, arise from the following simple mechanism. Imagine a population of stem cells each of which has, during a short time interval $[t, t + \mathrm{d}t]$, a probability $\lambda_b \mathrm{d}t$ of producing a copy of itself (a birth) and a probability $\lambda_d \mathrm{d}t$ of losing itself (cell death). The coefficients λ_b and λ_d can be interpreted as the respective transition rates of birth and death of an individual. If there are n individuals present in the colony at time t, it follows, from the additive rule of probabilities of mutually exclusive events, that during the short time interval, the probability of a birth somewhere in the colony is $\lambda_b n \mathrm{d}t$, whereas the probability of a death is $\lambda_d n \mathrm{d}t$. In the next chapter, we will learn how to turn this information of birth/death probability into a differential equation for the probability $P(n, t)$ of finding the colony in state n at time t.

4.6.3 Toward Markov Processes

In a general stochastic process, the probability of moving from one state to another can depend on its state at present time and past times. However, many stochastic processes rapidly forget their past history. In that case, it is reasonable to assume that the probability of moving from one state to another is entirely determined by the most recent state and the states at earlier times have no influence. This is the Markov assumption, and the process is called Markov process. In the next chapter we will return to Markov processes for a more formal discussion.

Problems

4.1. A substrate and an inhibitor are competing for a binding site on an enzyme. Assume that the probability that in a particular scenario, the substrate wins is 0.7. If 10 observations are made to see which of the two is bound to the enzyme, compute the probability that the substrate is bound in exactly five observations.

4.2. The color of one's eyes is determined, in a simplified model, by a single pair of genes, with the gene for brown eyes being dominant over the one for blue eyes. This means that an individual having two blue-eyed genes will have blue eyes, while one having either two brown-eyed genes or one brown-eyed and one blue-eyed gene will have brown eyes. When two people mate, the resulting offspring receives one randomly chosen gene from each of its parents' gene pair. If the eldest child of a pair of brown-eyed parents has blue eyes, what is the probability that exactly two of the four other children (none of whom is a twin) of this couple also have blue eyes?

4.3. Imagine a mechanism inside the cell that, during DNA replication, inspects 1000 DNA base pairs per minute. Assume that the probability of finding a replication error in a base pair is 0.001. What is the probability of finding at least three replication errors?

4.4. Red blood cells are replaced, on average, every 28 days. Assuming an exponential distribution for the lifetime of red blood cells, find the distribution of the time required for 10 turnovers (successive replacements).

4.5. Suppose that the life of a cell, measured in days, is exponentially distributed with an average value of 10 days. What is the probability that the cell will be intact after 5 days? If it turns our that the cell has survived 5 days, what is the probability that it survives another 5 days?

4.6. Write a Matlab code that generates a random number from the binomial distribution with (trials, success) parameters (n, p).

4.7. Write a Matlab code that generates a random number from the exponential distribution with parameter a. Extend the code to generate a random number from the gamma distribution with the (rate, shape) parameters (a, n).

4.8. Write a Matlab code that generates a random number from the geometric distribution with success probability p.

4.9. A stem cell divides into two daughter cells that may be either both stem cells, with probability $1/5$, or both committed progenitor cells, with probability $1/5$, or else as in invariant asymmetric division, one daughter cell may be stem cell (self renewal) and the other a progenitor (due to differentiation), with probability $3/5$. Define the random variable X to be the number of daughter cells that are stem cells. Write down the probability distribution for X. How would you generate samples of X?

4.10. Transcription factors are molecules that are activated by signaling pathways to regulate gene expression. The arrivals of transcription factors at a promoter region can be modeled as a Poisson process. Suppose that on average 30 molecules arrive per minute. What would you choose for the rate λ in arrivals per second? What is the expected interarrival time?

Chapter 5

Stochastic Modeling of Biochemical Networks

In this chapter, we present a stochastic framework for modeling subcellular biochemical reaction networks. In particular, we make an effort to show how the notion of propensity, the chemical master equation (CME), and the stochastic simulation algorithm arise as consequences of the Markov property. We would encourage the reader to pay attention to this, because it is not easy to see this connection when reading the relevant literature in systems biology. We review various analytical approximations of the CME, leaving out stochastic simulation approaches reviewed in [113, 155]. Moreover, we sketch interrelationships between various stochastic approaches. The books [114] and [165] inspired this chapter and can be referred to for further reading.

5.1 Stochastic Formulation

Since the occurrence of reactions involves discrete and random events at the microscopic level, it is impossible to deterministically predict the progress of reactions in terms of the macroscopic variables (observables) $N(t)$ and $Z(t)$. To account for this uncertainty, one of the observables $N(t), Z(t), X(t)$ is formulated as a stochastic process. If we choose the copy number $N(t)$ as the observable of the system, a sample value n of the process is the state of our biochemical system under consideration.

Our goal is to determine how the process $N(t)$ of copy numbers evolves in time. Starting at time $t = 0$ from some initial state $N(0)$, every sample path of the process remains in state $N(0)$ for a random amount of time W_1 until the occurrence of a reaction takes the process to a new state $N(W_1)$; it remains in state $N(W_1)$ for another random amount of time W_2 until the occurrence of another reaction takes the process to a new state $N(W_1 + W_2)$, and so on, as shown in Figure 5.1. In other words, the time-dependent copy number $N(t)$ is a *jump process*.

The stochastic process $N(t)$ is characterized by a collection of state

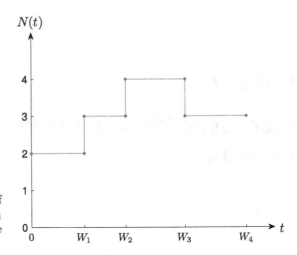

Figure 5.1
A time-course realization of a jump process. Labels W_i on the time axis denote the waiting times.

probabilities and transition probabilities. The state probability

$$P(n, t) = \Pr\Big[N(t) = n\Big]$$

is the probability that the process $N(t)$ is in state n at time t. The transition probability

$$\Pr\Big[N(t_0 + t) = n \,|\, N(t_0) = m\Big]$$

is the conditional probability that process $N(t)$ has moved from state m to state n during the time interval $[t_0, t_0 + t]$. The analysis of a stochastic process becomes greatly simplified when the above transition probability depends on (i) the starting state m but not on the states before time t_0 and (ii) the interval length t but not on the start time t_0. Property (i) is the well-known *Markov property*, and a process with this property is said to be a *Markov process*. The process holding property (ii) is said to be a *homogeneous process*. If the molecules are well mixed and are available everywhere for a reaction (space can be ignored), then the copy number $N(t)$ can be approximately formulated as a homogeneous Markov process in continuous time. In this text, all Markov processes will be assumed to be homogeneous unless stated otherwise. Now we use a simplified notation for the above transition probability:

$$\begin{aligned} P(n|m, t) &= \Pr\Big[N(t_0 + t) = n \,|\, N(t_0) = m\Big] \\ &= \Pr\Big[N(t) = n \,|\, N(0) = m\Big]. \end{aligned} \tag{5.1}$$

It should be remembered that t in the above equation is the length of the time interval. The initial condition is usually fixed and the state probability can be written as a transition probability

$$P(n, t) = P(n|n^0, t) = \Pr\left[N(t) = n \,|\, N(0) = n^0\right].$$

The Markov property has two important consequences, explained in the following two sections.

5.1.1 Chapman–Kolmogorov Equation

The Markov property places a consistency condition on the transition probabilities. To see this, decompose the transition probability, as

$$\Pr\left[X(t + w) = n \,|\, X(0) = m\right]$$
$$= \sum_{n'} \Pr\left[X(t + w) = n \,|\, X(t) = n' \cap X(0) = m\right] \Pr\left[X(t) = n' \,|\, X(0) = m\right]$$
$$= \sum_{n'} \Pr\left[X(t + w) = n \,|\, X(t) = n'\right] \Pr\left[X(t) = n' \,|\, X(0) = m\right]$$

where the Markov property allows us to ignore the information $X(0) = m$ (about the past event) in the joint condition (in the second line). In the compact notation for transition probabilities, the above consistency condition takes the form

$$P(n|m, t + w) = \sum_{n'} P(n|n', w) P(n'|m, t), \tag{5.2}$$

which is known as the *Chapman–Kolmogorov equation* (CKE) for continuous-time Markov processes. This equation expresses the probability of a transition $(m \to n)$ as the summation of probabilities of all transitions $(m \to n' \to n)$ via the intermediate states n'. Figure 5.2 illustrates the idea conveyed by the CKE. It is important to clarify that the CKE is only a consistency condition imposed on every stochastic process by the Markov property and cannot characterize a particular process. We need dependence relations between random variables of the process to characterize it. Typically, that is achieved by investigating the local behavior of transition probabilities in a short time interval. Replacing the length w of the time interval of the transition probabilities in (5.2) by Δt and fixing the initial condition, the CKE (5.2) reduces to

$$P(n, t + \Delta t) = \sum_{n'} P(n|n', \Delta t) P(n', t), \tag{5.3}$$

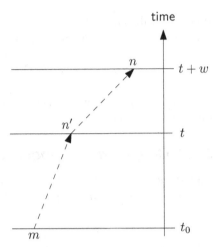

Figure 5.2
Graphical interpretation of the Chapman–
Kolmogorov equation. The probability of a
transition $m \to n$ can be obtained by summing up the probabilities of all transitions
$m \to n' \to n$, via intermediate states n'.
Drawing adopted from [54].

where the transition probabilities away from the fixed initial state have been
replaced by the state probabilities. Later we will see that the short-time
transition probabilities $P(n|n', \Delta t)$ can be expressed in terms of parameters of
the particular process under consideration when certain modeling assumptions
about the underlying chemical reactions are made. This will open the door
for an analytical characterization of a particular Markov process in Section
5.5.

5.1.2 The Memoryless Property

Suppose the Markov process $N(t)$ is in state n at time t_0 and let $T_j(n)$ denote
the time in state n until the occurrence of a reaction R_j takes the process
to state $n + S_{.j}$, where $S_{.j}$ is the jth column of the stoichiometry matrix
(Chapter 2). If the reaction has not occurred during $[t_0, t_0 + w]$, we can write
$T_j(n) > w$. This knowledge, however, does not change the uncertainty in time
until the next reaction. In other words, the process is *memoryless,* and its
subsequent behavior is independent of w. The mathematical representation
(4.9) of this memoryless property was discussed in the previous chapter, which
we rewrite here in the present notation as

$$\Pr\left[T_j(n) > w + t \,|\, T_j(n) > w\right] = \Pr\left[T_j(n) > t\right].$$

This holds only for the exponential distribution, as shown in the previous
chapter.

 The memoryless property, and hence the fact that the times between
reactions are exponentially distributed, opens the door for stochastic simula-

tions of biochemical reaction networks. That will be our focus in the following section.

5.2 Propensity as a Transition Rate

We here derive the basis for a stochastic simulation algorithm, which has played an important role in applications of stochastic modeling to systems biology [113, 155]. It follows from the previous section that the time $T_j(n)$ until the occurrence of reaction R_j has an exponential distribution with a parameter, say $a_j(n)$. We can thus write

$$\Pr\left[T_j(n) > t\right] = \exp\left(-a_j(n)t\right) \tag{5.4}$$

for the probability that an R_j reaction will not occur in the next time interval of length t, provided reactions of other types involving R_j reactants do not occur during this period. Using a Taylor expansion, for an arbitrarily short interval of length Δt, the above probability can be written as

$$\Pr\left[T_j(n) > \Delta t\right] = \exp\left(-a_j(n)\Delta t\right) = 1 - a_j(n)\Delta t + o(\Delta t), \tag{5.5}$$

where the sum of the second- and higher order terms has been written as $o(\Delta t)$ because it will vanish faster than Δt as the interval is made vanishingly short. The probability of occurrence of an R_j reaction during the same short interval is complementary to the above:

$$\Pr\left[T_j(n) \leq \Delta t\right] = a_j(n)\Delta t + o(\Delta t). \tag{5.6}$$

The rate parameter $a_j(n)$, which gives the probability per unit time of the occurrence of an R_j reaction in state n, is referred to as the *reaction propensity*. Although propensity is often is seen as a measure of how fast a reaction proceeds, it is important to remember it as a "probability per unit time" in contrast to the conversion rate $\hat{v}_j(n)$, which is the number of R_j occurrences per unit time. In other words, while $a_j(n)\Delta t$ gives the probability of an R_j occurrence in a short time interval of length Δt, the number of R_j occurrences in the same interval is given by $\hat{v}_j(n)\Delta t$. Recall that the propensity $a_j(n)$ is defined at a particular state n and is hence a deterministic quantity. However, it can be interpreted as a sample value of a random variable $a_j(N(t))$ that is a function of the random variable $N(t)$, the copy number at a fixed time t. If t is varying, then $N(t)$, and hence $a_j(N(t))$, is a stochastic process.

5.2.1 Mean Conversion Rate

In a vanishingly short interval, it is highly improbable that a particular reaction will occur more than once. To see this, note that the probability of two occurrences of R_j during a time interval $[t, t + \Delta t]$ is the joint probability of its first occurrence during $[t, t + \alpha \Delta t]$ and a second occurrence during $(t + \alpha \Delta t, t + \Delta t]$:

$$
\Pr\Big[T_j(n) \le \alpha \Delta t\Big] \Pr\Big[T_j(n + S_{.j}) \le (1 - \alpha)\Delta t\Big]
$$
$$
= \Big(a_j(n)\alpha \Delta t + o(\Delta t)\Big)\Big(a_j(n + S_{.j})(1 - \alpha)\Delta t + o(\Delta t)\Big) = o(\Delta t),
$$

where $0 < \alpha < 1$. Therefore, the probability in (5.6) is equivalent to the probability in state n of *one* reaction (i.e., a unit increment in the reaction count) of type R_j during $[t, t + \Delta t]$:

$$
\Pr\Big[Z_j(t + \Delta t) - Z_j(t) = 1 \,|\, N(t) = n\Big] = a_j(n)\Delta t + o(\Delta t).
$$

The probability distribution in state n of the short-time R_j reaction-count increment $\Delta Z_j = Z_j(t + \Delta t) - Z_j(t)$ during the time interval $[t, t + \Delta t)$ is

$$
\Pr\Big[\Delta Z_j = z_j \,|\, N(t) = n\Big] = o(\Delta t) + \begin{cases} a_j(n)\Delta t & \text{if } z_j = 1, \\ 1 - a_j(n)\Delta t & \text{if } z_j = 0, \\ 0 & \text{if } z_j > 1. \end{cases} \tag{5.7}
$$

It is interesting to note that for a vanishingly small Δt, the change in propensity during the interval is vanishingly small, which in turn implies that the above distribution approaches a Poisson distribution with mean (and variance) $a_j(n)\Delta t$. The expected value, conditioned on $N(t) = n$, of this short-time R_j reaction-count increment is a summation over all possible values z_j of the reaction-count increment, each weighted by its probability,

$$
\langle \Delta Z_j \rangle_n = \langle \Delta Z_j \,|\, N(t) = n \rangle
$$
$$
= \sum_{z_j=0}^{\infty} z_j \Pr\Big[\Delta Z_j = z_j \,|\, N(t) = n\Big]
$$
$$
= \overbrace{a_j(n)\Delta t}^{z_j=1} + \overbrace{o(\Delta t)}^{z_j>1}. \tag{5.8}
$$

The subscript n here reflects conditioning on $N(t) = n$. Comparing the above result with the rate law (2.15), the reaction propensity $a_j(n)$ can be interpreted as the (conditional) *mean conversion rate* from state n. The unconditional expectation of the short-time R_j reaction-count increment can be obtained by summing the probabilities $P(n,t)$ weighted by the above conditional expectation over all possible states n:

$$\left\langle \Delta Z_j \right\rangle = \sum_n \left\langle \Delta Z_j \right\rangle_n P(n,t)$$

$$= \sum_n a_j(n)P(n,t)\Delta t + o(\Delta t)$$

$$= \left\langle a_j\big(N(t)\big) \right\rangle \Delta t + o(\Delta t),$$

which for vanishingly small Δt leads to the ODE

$$\frac{\mathrm{d}}{\mathrm{d}t}\langle Z_j(t)\rangle = \left\langle a_j\big(N(t)\big)\right\rangle . \tag{5.9}$$

This ODE conveys an important relation: the expected reaction count increases with time at a rate equal to the expected reaction propensity. In other words, analogous to (2.15), the mean conversion rate is equal to the mean reaction propensity.

5.2.2 Reaction as a State Transition

The state transition from state n associated with channel R_j will be written as

$$n \xrightarrow{\;\;a_j(n)\;\;} n + S_{\bullet j},$$

because the completion of one R_j reaction simply adds the jth column of the stoichiometry matrix to the state. The completion of one R_j reaction could also bring the system into state n from another state. This state transition can be written as

$$n - S_{\bullet j} \xrightarrow{\;\;a_j\left(n - S_{\bullet j}\right)\;\;} n .$$

5.2.3 Propensity Rate Law

The dependence relation of the propensity on the state n is determined by the system being modeled and reflects the assumptions made about the system. If R_j is an elementary reaction in a well-mixed system, it is reasonable to assume that each possible combination of the R_j reactant molecules has the same probability c_j per unit time to react. In other words, $c_j \mathrm{d}t$ gives the probability that a particular combination of R_j reactant molecules will react in a short time interval $(t, t + \mathrm{d}t]$. In the literature, c_j is referred to as the *stochastic (reaction) rate constant*. If there are $h_j(n)$ different possible combinations of

R_j reactant molecules in state n, then the propensity $a_j(n)$ can be written as

$$a_j(n) = c_j h_j(n).$$ (5.10)

The form of $h_j(n)$ depends on the order of the reaction R_j.

Zero-order reaction ($\varnothing \to X$): Since the reaction rate does not depend on the reactant, the propensity is a constant $a(n) = c_j$ if the reaction is a single step. If the reaction is nonelementary, then the propensity is a function of the copy numbers of any enzymatic chemical species involved.

First-order reaction ($X \to$): The stochastic reaction rate c_j of this reaction is the probability per unit time of a particular reactant molecule undergoing the reaction. Given n reactant molecules, the probability per unit time of any reactant molecule undergoing the reaction is obtained by summing the individual probabilities of all n reactant molecules, that is, $a_j(n) = c_j n$.

Bimolecular reaction ($X_1 + X_2 \to$): The stochastic reaction rate c_j of this reaction is the probability per unit time of a particular pair of reactant molecules undergoing the reaction. Given n_1 copies of reactant X_1 and n_2 copies of reactant X_2, there are $n_1 n_2$ distinct possible pairs of reactant molecules available for the reaction. The probability per unit time of any pair of reactant molecules undergoing the reaction is obtained by summing the individual probabilities of all $n_1 n_2$ pairs of reactant molecules, that is, $a_j(n) = c_j n_1 n_2$.

Bimolecular reaction ($2X_1 \to$): The stochastic reaction rate c_j of this reaction is the probability per unit time of a particular pair of reactant molecules undergoing the reaction. Given n copies of reactant X, there are $(n-1)n/2$ distinct possible pairs of reactant molecules available for the reaction. The reaction propensity is thus $a_j(n) = c_j (n-1)n/2$.

For an elementary reaction channel R_j of the general form (2.1), with $\underline{S}_{ij}$ molecules of reactant species X_i, we can write the combinatorial function

$$h_j(n) = \prod_{i=1}^{s} \binom{n_i}{\underline{S}_{ij}}.$$ (5.11)

However, it is highly unlikely that a reaction of order higher than two will result from all its reactants coming together and reacting in one step, for example by collision. A more realistic model will decompose the high-order reaction into two or more one-step reactions.

M-code 5.1 `makePropensity`: implements propensity function (5.10).

```
function a = makePropensity(c,Su)
r = size(Su,2);
inz = (Su>0);
a = @RateLaw;
    function w = RateLaw(n)
        C = repmat(n,1,r);
        C(~inz) = 1;
        ineg = (C<Su);
        C(inz & ineg) = 0;
        i = (inz & ~ineg);
        C(i) = arrayfun(@nchoosek,C(i),Su(i));
        w = c.*prod(C)';
    end
end
```

Matlab implementation: When all the reaction channels are elementary, the above combinatorial expression determines the number $h_j(n)$ of reactant combinations that multiplies the rate constant c_j in (5.10) for the reaction propensities $a_j(n)$. To implement such a computation in Matlab, let us represent the number r of reaction channels by Matlab scalar r, the matrix S (of stoichiometries on the left) by Matlab matrix Su and the stochastic rate constant c by Matlab vector c. Then the propensity function $a(\cdot)$ can be implemented as a function handle a in the following Matlab code:

```
a = @(n) c.*prod(arrayfun(nchoosek, repmat(n,1,r), Su))';
```

where the Matlab function `nchoosek` has been employed to implement the combinatorial computation of (5.11). As pointed out before, the compact code above is not efficient for large reaction networks because of the computationally expensive combinatorics and multiplications involved. Another, more serious, issue with the code is that the function `nchoosek` does not allow its second argument to be greater than its first one. This latter case arises when the abundance of a reactant species decreases below its participation in a reaction channel. To avoid the aforementioned issues and any unnecessary computations, the code is replaced by the function `makePropensity` in M-code (5.1). Here the output a returned by the main function `makePropensity` is a function handle to the nested function `Propensity`. Note how extra

Table 5.1 Examples of elementary reactions R_j listed with their propensities $a_j(n)$ and conversion rates $\hat{v}_j(n)$. The last column shows the condition for equality of the two. The resulting relationship between c_j and $\hat{k}_j$ is translated in terms of the rate constant k_j following (2.1). Note that cases of repeating reactant species require a large system size, $\Omega \gg 1$.

R_j	$a_j(n)$	$\hat{v}_j(n)$	$a_j(n) = \hat{v}_j(n)$ if
$\emptyset \xrightarrow{k_j} X$	c_j	$\hat{k}_j$	$c_j = \hat{k}_j = \Omega k_j$
$X \xrightarrow{k_j} ?$	$c_j n$	$\hat{k}_j n$	$c_j = \hat{k}_j = k_j$
$X_1 + X_2 \xrightarrow{k_j} ?$	$c_j n_1 n_2$	$\hat{k}_j n_1 n_2$	$c_j = \hat{k}_j = \frac{k_j}{\Omega}$
$2X \xrightarrow{k_j} ?$	$c_j \frac{(n-1)n}{2}$	$\hat{k}_j n^2$	$c_j = 2\hat{k}_j = \frac{2k_j}{\Omega}$
$X_1 + X_2 + X_3 \xrightarrow{k_j} ?$	$c_j n_1 n_2 n_3$	$\hat{k}_j n_1 n_2 n_3$	$c_j = \hat{k}_j = \frac{k_j}{\Omega^2}$
$X_1 + 2X_2 \xrightarrow{k_j} ?$	$c_j n_1 \frac{(n_2-1)n_2}{2}$	$\hat{k}_j n_1 n_2^2$	$c_j = 2\hat{k}_j = \frac{2k_j}{\Omega^2}$

combinatorial computations are avoided for the obvious cases $\underline{S}_{ij} = 0$ and $\underline{S}_{ij} = 1$.

For elementary reactions, the stochastic rate constant c is closely related to the deterministic rate constant, as shown below.

5.2.4 Deterministic- and Stochastic Reaction Rates

Using the interpretation of propensity as the mean reaction count per unit time from (5.9), the propensity is analogous to the conversion rate $\hat{v}_j$ defined earlier in the deterministic framework. Hence, the propensity can be interpreted as the *stochastic conversion rate*. The two kinds of rates are given for selected elementary reactions in Table 5.1. The condition under which the two type of extensive reaction rates are equal is shown in the corresponding entry of the last column. This also provides the relationship between the stochastic rate constant c_j and the deterministic rate constant k_j. That relationship is generalized in the following section. Note that the two rates differ by an order of one molecule in the case of repeating reactant species:

$$a_j(n) = \hat{v}_j(n) + \mathcal{O}(1)$$

Here $\mathcal{O}(x)$ represents a function (neglected term) bounded by some linear function of x. The difference $\mathcal{O}(1)$ will vanish for an infinitely large system. The above relation can also be translated to a relation between propensity and reaction rate, namely

$$a_j(n) = \Omega\left(v_j(x) + \mathcal{O}\left(\Omega^{-1}\right)\right), \tag{5.12}$$

where $n = \Omega x$ is the assumed system state [33].

5.2.5 Deterministic- and Stochastic Rate Constants

Let us find the conditions under which the deterministic extensive reaction rate and propensity of a general elementary reaction are approximately the same. From (2.18), (5.10), and (5.11) we can propose

$$\hat{k}_j \prod_{i=1}^{s} n_i^{\underline{S}_{ij}} = \hat{v}_j(n) \approx a_j(n) = c_j \prod_{i=1}^{s} \binom{n_i}{\underline{S}_{ij}}.$$

The leftmost expression is valid only in the deterministic framework, which requires large system size, $\Omega \gg 1$. To the extent that this assumption is valid, the combinatorial function can be approximated as

$$\binom{n_i}{\underline{S}_{ij}} = \frac{(n_i - \underline{S}_{ij} + 1) \cdots (n_i - 1) n_i}{\underline{S}_{ij}!}$$

$$= \left(\frac{\Omega^{\underline{S}_{ij}}}{\underline{S}_{ij}!}\right)\left(x_i - \frac{\underline{S}_{ij} - 1}{\Omega}\right) \cdots \left(x_i - \frac{1}{\Omega}\right) x_i$$

$$\approx \left(\frac{\Omega^{\underline{S}_{ij}}}{\underline{S}_{ij}!}\right) x_i^{\underline{S}_{ij}} \quad \text{for} \quad \Omega \gg 1.$$

Inserting this into the previous equation leads to the stochastic rate constant

$$c_j = \hat{k}_j \prod_{i=1}^{s} (\underline{S}_{ij}!) = \frac{k_j}{\Omega^{K_j - 1}} \prod_{i=1}^{s} (\underline{S}_{ij}!), \tag{5.13}$$

where $K_j = \sum_{i=1}^{s} \underline{S}_{ij}$ is the number of R_j reactant molecules required to collide and possibly result in a single occurrence of the reaction. The above derivation is a refinement of our earlier attempt in [169].

Example 5.1 (Standard modification) In the standard modification (2.5), the copy number $N(t)$ of the unmodified proteins is a simple birth–death process. Each copy of the unmodified protein U is modified at a rate k_w. Similarly, each copy of the modified protein W is demodified at a rate k_u. Both the modification and the demodification are monomolecular reactions. With $0 < n < n^{\text{tot}}$ unmodified proteins, expressions for the reaction propensities $a(n)$ are listed here (on the right) together with the corresponding reactions on the left:

$$\left.\begin{aligned}
\text{U} \xrightarrow{k_w} \text{W}, & \qquad a_w(n) = k_w n, \\
\text{W} \xrightarrow{k_u} \text{U}, & \qquad a_u(n) = (n^{\text{tot}} - n)k_u.
\end{aligned}\right\} \tag{5.14}$$

Example 5.2 (Heterodimerization) The reversible heterodimerization (2.6) is a 3-component 2-reaction network. Let $N_1(t)$, $N_2(t)$, and $N_3(t)$ denote, the respective copy numbers of the components X_1, X_2, and X_3. The full state has to respect the two conservation relations (2.22), which translate to

$$N_1(t) + N_3(t) = \hat{q}_1 \quad \text{and} \quad N_2(t) + N_3(t) = \hat{q}_2,$$

where $\hat{q}_1 = \Omega q_1$ and $\hat{q}_2 = \Omega q_2$ are the conserved copy numbers and $\Omega = N_A V$ is the system size. The Markov process $N(t) = N_3(t)$ having states $n = n_3$ is sufficient to describe the system, because the remaining two variables can be determined from the conservation relations above. Subject to those conservation relations, expressions for the channel propensities $a(n)$ in state $n = n_3$ are listed here (on the right) together with the corresponding reactions on the left:

$$\left.\begin{aligned}
\text{X}_1 + \text{X}_2 \xrightarrow{k_1} \text{X}_3, & \qquad a_1(n) = \hat{k}_1 (\hat{q}_1 - n)(\hat{q}_2 - n), \\
\text{X}_3 \xrightarrow{k_2} \text{X}_1 + \text{X}_2, & \qquad a_2(n) = k_2 n.
\end{aligned}\right\} \tag{5.15}$$

Example 5.3 (Lotka–Volterra model) The mutual interaction between two kinds of entities depicted in (2.7) is a 2-component 3-reaction network. Let $N_1(t)$ denote the population of the prey X_1, and $N_2(t)$ that of the predator X_2. The prey replication and the predation are of the second order, whereas predator death is of first order. Expressions for the channel propensities $a(n)$ in state $n = (n_1, n_2)^T$ are listed here (on the right) together with the corresponding reactions on the left:

$$\left.\begin{aligned}
\text{X}_1 + \text{A} \xrightarrow{\hat{k}_1} 2\text{X}_1, & \qquad a_1(n) = \hat{k}_1 n_A n_1, \\
\text{X}_1 + \text{X}_2 \xrightarrow{\hat{k}_2} 2\text{X}_2, & \qquad a_2(n) = \hat{k}_2 n_1 n_2, \\
\text{X}_2 \xrightarrow{\hat{k}_3} \varnothing, & \qquad a_3(n) = \hat{k}_3 n_2.
\end{aligned}\right\} \tag{5.16}$$

Example 5.4 (Enzyme kinetic reaction) The enzyme kinetic model (2.8) is a 4-component 3-reaction network. Let $N_E(t)$ denote the copy number of the enzyme, $N_S(t)$ that of the substrate, $N_{ES}(t)$ that of the complex, and $N_P(t)$ that of the product. The full state has to respect the two conservation relations (2.24), which translate to

$$N_E(t) + N_{ES}(t) = n_E^{tot} \quad \text{and} \quad N_S(t) + N_{ES}(t) + N_P(t) = n_S^{tot},$$

where $n_E^{tot} = \Omega x_E^{tot}$ and $n_S^{tot} = \Omega x_S^{tot}$ are the conserved copy numbers and $\Omega = N_A V$. The Markov process

$$N(t) = \left(N_S(t), N_{ES}(t)\right)^T$$

having states $n = (n_S, n_{ES})^T$ is sufficient to describe the system, because the remaining two variables can be determined from the conservation relations above. The (enzyme–substrate) complex formation is a bimolecular reaction, whereas the complex dissociation and the product formation are monomolecular reactions. Expressions for the reaction propensities $a(n)$ in state $n = (n_S, n_{ES})^T$ are listed here (on the right) together with the corresponding reactions on the left:

$$
\begin{array}{ll}
\text{E} + \text{S} \xrightarrow{k_1} \text{ES}, & a_1(n) = \hat{k}_1 \left(n_E^{tot} - n_{ES}\right) n_S, \\[2mm]
\text{ES} \xrightarrow{k_2} \text{E} + \text{S}, & a_2(n) = k_2 n_{ES}, \\[2mm]
\text{ES} \xrightarrow{k_3} \text{E} + \text{P}, & a_3(n) = k_3 n_{ES}.
\end{array}
\right\} \quad (5.17)
$$

Example 5.5 (Schlögl model) For the Schlögl reaction scheme (2.9), let x_A and x_B denote the constant respective concentrations of chemicals A and B, and let $N(t)$ denote the time-dependent copy number of chemical X. The first two reaction channels, the autocatalysis and its backward dissociation, are trimolecular reactions with two and three identical species, respectively. The last two reaction channels, the synthesis/dissociation of X from/to B, are monomolecular reactions. Expressions for the reaction propensities $a(n)$ in state $n = (n_1, n_2)^T$ are listed here (on the right) together with the corresponding reactions on the left:

$$
\begin{array}{ll}
\text{A} + 2\text{X} \xrightarrow{k_1} 3\text{X}, & a_1(n) = \hat{k}_1 n \left(n - 1\right), \\[2mm]
3\text{X} \xrightarrow{k_2} \text{A} + 2\text{X}, & a_2(n) = \hat{k}_2 n \left(n - 1\right)\left(n - 2\right), \\[2mm]
\text{B} \xrightarrow{k_3} \text{X}, & a_3(n) = \hat{k}_3, \\[2mm]
\text{X} \xrightarrow{k_4} \text{B}, & a_4(n) = k_4 n,
\end{array}
\right\} \quad (5.18)
$$

where the new rate parameters are defined as

$$\hat{k}_1 = \frac{k_1 x_A}{\Omega}, \quad \hat{k}_2 = \frac{k_2}{\Omega^2}, \quad \hat{k}_3 = k_3 x_B \Omega,$$

in terms of the system size $\Omega = N_A V$. Note that the constant concentrations have been digested in the conversion rate constants for a simpler notation.

Network reduction: The last example shows an interesting feature of some biochemical reaction networks. In this example, reaction channels R_1 and R_3 both have the same stoichiometry as far as the abundance $N(t)$, the only state variable, is concerned. A reaction of either of the two channels will take the system from state n to state $n+1$. Similarly, a reaction of either channel R_1 or R_4 will take the system from state n to state $n-1$. Thus, as far the state transitions are concerned, the reaction network (5.18) can be reduced to a birth–death process with birth rate $a^+(n)$ and death rate $a^-(n)$ given by

$$\left. \begin{array}{l} a^+(n) = \hat{k}_1 n\,(n-1) + \hat{k}_3, \\[2mm] a^-(n) = \hat{k}_2 n\,(n-1)\,(n-2) + k_4 n\,. \end{array} \right\} \tag{5.19}$$

In general, if the stoichiometry matrix S of a reaction network has identical columns, the network can be reduced by merging the set of reaction channels corresponding to those columns in the above manner.

Example 5.6 (Gene regulation) For the gene regulation scheme (2.11) write $n_M(t)$, $n_G(t)$, and $n_P(t)$ for the respective time-dependent copy numbers of mRNA M, the unbound gene G, and protein P. The total gene copy number n_G^{tot} is assumed to be constant, so that the bound (repressed) protein concentration is simply $n_G^{tot} - n_G$. The reaction propensities based on mass-action kinetics are (each to the right of the corresponding reaction channel)

$$\left. \begin{array}{ll} G \xrightarrow{\ k_m\ } G + M, & a_m(n) = k_m n_G, \\[2mm] M \xrightarrow{\ k_p\ } M + P, & a_p(n) = k_p n_M, \\[2mm] G + P \xrightarrow{\ k_b\ } GP, & a_b(n) = k_b n_G n_P, \\[2mm] GP \xrightarrow{\ k_u\ } G + P, & a_u(n) = k_u \left(n_G^{tot} - n_G \right), \\[2mm] M \xrightarrow{\ k_m^-\ } \varnothing, & a_m^-(n) = k_m^- n_M, \\[2mm] P \xrightarrow{\ k_p^-\ } \varnothing, & a_p^-(n) = k_p^- n_P\,. \end{array} \right\} \tag{5.20}$$

Stochastic simulation: The idea of reaction propensity allows one to derive strategies for generating sample paths of the stochastic process $N(t)$. The practice of generating sample paths of a stochastic process is referred to as *stochastic simulation*. In the following two sections we look at two alternatives: discrete and continuous stochastic simulation.

5.3 Discrete Stochastic Simulation

The discrete approach to stochastic simulation decomposes the problem by asking two successive questions: (1) *when* is the next reaction going to occur and (2) *what* type of reaction will it be?

5.3.1 Time Until the Next Reaction

Suppose that a process is in state n. The time $T_0(n)$ until the next reaction is a continuous random variable, which can be interpreted as the *exit time* (of the process away) from state n. It turns out that the exit time is also exponentially distributed. To see this, we consider the probability that no reaction has occurred in an interval of length t. Divide the interval into a large number K of subintervals, each of length $\Delta t = t/K$, so short that at most one reaction can occur in a subinterval, with probability $a_j(n)\Delta t$. The required probability then from independence of nonoverlapping intervals is

$$
\Pr\left[T_0(n) > t\right] = \lim_{K \to \infty} \left[\prod_j \left(1 - a_j(n)\frac{t}{K}\right)\right]^K
$$

$$
= \lim_{K \to \infty} \prod_j \left(1 - a_j(n)\frac{t}{K}\right)^K
$$

$$
= \prod_j \exp\left(-a_j(n)t\right) = \exp\left(-t\sum_j a_j(n)\right).
$$

Hence the exit time $T_0(n)$ from state n is exponential with rate parameter

$$
a_0(n) = \sum_j a_j(n),
$$

which is the *exit rate* (of the process away) from state n. The mean exit time is simply the reciprocal $1/a_0(n)$, which allows an interesting observation. Since the $a_0(n)$ will increase/decrease with an increase/decrease in the copy numbers of the reactant species, the mean exit time will increase/decrease accordingly. In other words, large/small copy numbers of the reactant species lead to frequent/rare reactions. Recall from Figure 1.5 that only frequent reactions allow for continuous approximation of an inherently discrete process.

Random number generation: The exponential time $T_0(n)$ has CCDF

$$G(t) = \Pr\Big[T_0(n) > t\Big] = \exp\Big(-a_0(n)t\Big).$$

If u_1 is a uniform random number from $[0, 1]$, then following (4.10),

$$\tau = G^{-1}(u_1) = -\frac{\log u_1}{a_0(n)} \tag{5.21}$$

is a sample of the time until the next reaction.

5.3.2 Index of the Next Reaction

If it is known that a reaction has occurred in state n, the (conditional) probability that it was an R_j reaction is determined as

$$\lim_{\Delta t \to 0} \Pr\Big[T_j(n) \le \Delta t \,|\, T_0(n) \le \Delta t\Big]$$

$$= \lim_{\Delta t \to 0} \frac{\Pr\Big[T_j(n) \le \Delta t\Big]}{\Pr\Big[T_0(n) \le \Delta t\Big]} = \lim_{\Delta t \to 0} \frac{a_j(n)\Delta t + o(\Delta t)}{a_0(n)\Delta t + o(\Delta t)} = \frac{a_j(n)}{a_0(n)}.$$

Thus the index $J(n)$ of the next reaction known to have occurred in state n is a discrete random variable taking values j with probability

$$\Pr\Big[J(n) = j\Big] = \frac{a_j(n)}{a_0(n)}. \tag{5.22}$$

Random number generation: The index $J(n)$ has CCDF

$$F(j) = \Pr\Big[J(n) \le j\Big] = \sum_{l=1}^{j} \frac{a_l(n)}{a_0(n)}.$$

If u_2 is a uniform random number from $[0, 1]$ then, following (4.11),

$$j = F^{-1}(u_2) = \min_{w} \{w : F(w) \ge u_2\}$$

is a sample of the random index $J(n)$. For the range of values taken by J, the above condition is equivalent to

$$F(j-1) < u_2 \leq F(j).$$

Multiplying both sides by $a_0(n)$ and plugging in values for $F(j)$ gives the criteria

$$\sum_{l=1}^{j-1} a_l(n) < u_2 a_0(n) \leq \sum_{l=1}^{j} a_l(n). \tag{5.23}$$

for j to be a sample of the index $J(n)$ of the next reaction known to have occurred in state n.

5.3.3 Gillespie Algorithm

The two results (5.21) and (5.23) allow a simple procedure to simulate the Markov process: (1) Pick a sample τ from the exponential distribution with rate $a_0(n)$ to compute the time until the next reaction will occur, and (2) pick a sample j from the discrete distribution with probabilities (5.22) to determine the type of the next reaction. This is the stochastic simulation algorithm (SSA), known as the "Gillespie algorithm" [52] and involves the following steps:

1. Initialize the system at $t = 0$ with initial numbers of molecules for each species, $n_1, \ldots, n_s$.

2. For each $j = 1, \ldots, r$, calculate $a_j(n)$ based on the current state n.

3. Calculate the exit rate $a_0(n) = \sum_{j=1}^{r} a_j(n)$. Terminate if $a_0(n) = 0$.

4. Compute a sample τ of the time until the next reaction using (5.21).

5. Update the time $t = t + \tau$.

6. Compute a sample j of the reaction index using (5.23).

7. Update the state n according to R_j. That is, set $n = n + S_{.j}$, where $S_{.j}$ denotes jth column of the stoichiometry matrix S.

8. If $t < t_{\max}$, return to Step 2.

Improvements and approximations: For large biochemical systems, with many species and reactions, stochastic simulations (based on the original Gillespie algorithm) become computationally demanding. Recent years have seen a large interest in improving the efficiency/speed of stochastic simulations by modification/approximation of the original Gillespie algorithm. These improvements include the "next reaction" method [50], the "τ-leap" method [57] and its various improvements [20–22] and generalizations [23, 90], and the "maximal time step method" [124], which combines the next reaction and the τ-leap methods.

Matlab implementation: To implement the above SSA in Matlab, we need a representation of reactions and species. All the information about our chemical reaction network is encoded in the stoichiometry matrix (the static information) S and the reaction propensity (the kinetic information) $a(n)$ as a function of the state n. So all we need is a Matlab matrix S for the stoichiometry matrix S and a Matlab function handle a, when given the state n as an argument, for the propensity function $a(n)$. For elementary reactions, the function makePropensity in M-code 5.1 returns the required function handle representation of the propensity. If we pass these two arguments to the function makeSSA in M-code 5.2, a function handle ssa is returned that can be used as a function to generate sample trajectories according to the Gillespie SSA for the given chemical reaction network. In addition, a second function handle ensemb is returned that can be used to generate an ensemble of trajectories by executing multiple runs of SSA.

Software implementations: The Matlab implementation above is only for illustration and is by no means a practical one. An efficient implementation of the SSA and its variants is available in the Matlab *SimBiology toolbox* in the form of a stochastic solver. Numerous stochastic simulation packages, implemented in other programming languages, have been developed over time, including [15, 17, 49, 81, 129, 130]. An alternative software, which allows arbitrary rate laws (and hence nonelementary reactions), is Cains http://cain.sourceforge.net, a free tool for computationally efficient stochastic simulations.

Example 5.7 (Standard modification) To use the function handle returned by the function makeSSA in M-code 5.2, we need to specify the arguments S and a in the M-code 5.2. For the isomerization reaction (2.5) with propensities in (5.14), the two arguments are specified in the following piece of code:

```
S = [-1 1]; % stoichiometry matrix
k = [2;2]; % rate constant
a = @(n)[k(1)*n; k(2)*(ntot-n)]; % propensity
[ssa,ensem] = makeSSA(S,a);
n0 = 20; % initial condition
tmax = 5; dt = 0.01; % time scale and steps
[tt,nn] = ssa(n0,tmax,dt); % single SSA run
[TT,NN] = ensem(n0,tmax,dt,50)% 50 runs
```

M-code 5.2 makeSSA: implementation of Gillespie's SSA.

```
function [ssa,ensem] = makeSSA(S,a)
rng('shuffle'); s = size(S,1);
ssa = @gillespie; ensem = @ensemble;
    function [tt,nn] = gillespie(n,tmax,dt) % Single run
        t = 0; steps = tmax/dt;
        tt = zeros(steps,1);
        nn = zeros(steps,s);
        nn(1,:) = n; idx = 1;
        while t<tmax
            if all(n==0) % exhaustion check
                disp('Reactants_exhausted!');
                break;
            end
            asum = cumsum(a(n));
            t = t - (1/asum(end))*log(rand);
            j = find(asum>asum(end)*rand,1);
            n = n + S(:,j);
            if (t - tt(idx)) > dt
                idx = idx + 1;
                tt(idx) = t;
                nn(idx,:) = n;
            end
        end
        tt(idx:end) = []; nn(idx:end,:) = [];
    end
    function [TT,NN] = ensemble(n0,tmax,dt,runs) % Ensemble
        TT = (0:dt:tmax)'; ttmax = zeros(runs,1);
        NN = zeros(1+tmax/dt, s, runs);
        for i=1:runs
            [tt,nn] = gillespie(n0,tmax,dt);
            ttmax(i) = tt(end);
            NN(:,:,i) = interp1q(tt, nn, TT);
        end
        idx = (TT > min(ttmax));
        TT(idx) = []; NN(idx,:,:) = [];
    end
end
```

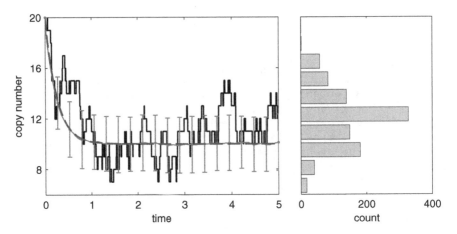

Figure 5.3 Stochastic simulation for the standard modification. *Left*: A single run of the SSA, mean over 1000 runs together with mean±SD (red thread), and solution of the deterministic ODE model (dashed). *Right*: Endpoint histogram. Parameters: $k_w = k_u = 2 \sec^{-1}$. Initial conditions: $n = 20$.

We note that the propensity function is specified as a function handle, unlike the stoichiometry field, which is a matrix. Another point to note is that the state variable n here is a scalar that is the copy number of unmodified proteins because the copy number of modified proteins is just ntot-n. Of course, values of k and ntot respectively corresponding to the rate constant vector k and the total copy number n^{tot} must be available in the Matlab workspace. The stochastic simulation results for the 2-species, 2-reaction network (2.5), with propensities (5.14), are shown in Figure 5.3. In the introductory chapter we briefly discussed the notion of identifiability of parameters from time-course data by taking the isomerization reaction as an example. To demonstrate the idea through stochastic simulation, see the four cases in Figure 5.4, wherein five sample trajectories are plotted, together with the associated deterministic time course, for each parameter-value pair.

We see different patterns as the difference $k_w - k_u$ of parameters is changed while keeping the sum $k_w + k_u$ the same. As will be seen in Chapter 6, the sum $k_w + k_u$ determines the mean trajectory, whereas the difference $k_w - k_u$ determines the spread of trajectories around the mean. The time-course measurements of mean alone provide information about the one fraction $k_u/(k_w+k_u)$ only. To get information about the other fraction $(k_w-k_u)/(k_w+k_u)$, we need time-course measurements of the variance as well.

Example 5.8 (Heterodimerization) The stochastic simulation results for the

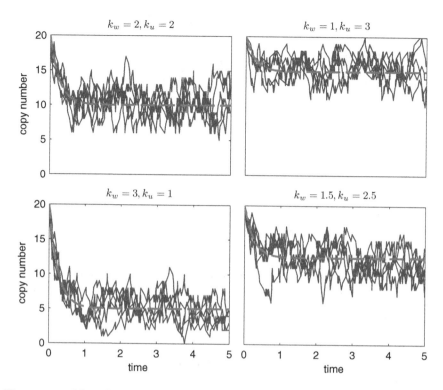

Figure 5.4 Identifiability illustrated through stochastic simulation. Five sample trajectories are shown, together with the associated deterministic time course, for each parameter-value pair. The parameter pairs (k_w, k_u) in sec^{-1} have been selected to satisfy $k_w + k_u = 4$. The total number of protein molecules was chosen to be $n^{\text{tot}} = 10$, initially all unmodified, that is, $N(0) = 10$.

reversible heterodimerization (2.6), with propensities (5.15), are shown in Figure 5.5. Notice the slight difference between the deterministic and the mean trajectories. This difference arises from the nonlinearity of the propensity of the bimolecular reaction.

Example 5.9 (Lotka–Volterra model) For the Lotka–Volterra system (2.7) with propensities in (5.16), the arguments S and a of the function `makeSSA` in the M-code 5.2 are specified in the following piece of code:

```
S = [1,-1,0; 0,1,-1];
a = @(n)[k(1)*nA*n(1); k(2)*n(1)*n(2); k(3)*n(2)];
```

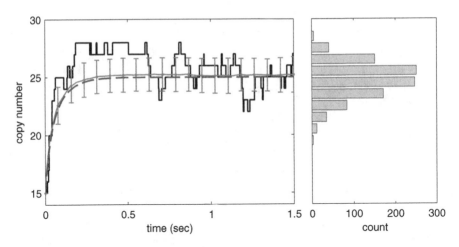

Figure 5.5 Stochastic simulation for heterodimerization. *Left*: A single run of the SSA (stairs), mean over 1000 runs together with mean±SD, and solution of the deterministic ODE model (dashed). *Right*: Endpoint histogram. Parameters: $k_1 = 1\,\mathrm{sec}^{-1}\,(\mathrm{nM})^{-1}$, $k_2 = 1\,\mathrm{sec}^{-1}$, $V = 1.66\,\mathrm{fL}$ (chosen so that $\Omega = 1\,(\mathrm{nM})^{-1}$), $q_1 = q_2 = 30\,\mathrm{nL}$. Initial concentrations: $x_1 = x_2 = x_3 = 15\,\mathrm{nM}$.

with the values of the variables k and nA corresponding respectively to the rate constant vector k and the constant copy number n_A, available in the Matlab workspace. Five sample trajectories are shown in Figure 5.6 side by side with the associated phase plot. To see the possibility of species extinction, sample trajectories starting from different initial populations are plotted in Figure 5.7. It can be seen that for some initial populations, species extinction occurs quickly.

Example 5.10 (Enzyme kinetic reaction) For the 4-species, 3-reaction enzymatic reaction (2.8) with propensities in (5.17), the arguments S and a of the function makeSSA in M-code 5.2 are specified in the following piece of code:

```
S = [-1,1,0; 1,-1,-1];
a = @(n)[c(1)*n(1)*(nStot-n(2)); c(2)*n(2); c(3)*n(2)];
```

which assumes that the values of the variables c and nStot corresponding respectively to the stochastic rate constant vector c and the total copy number n_S^{tot} of molecules involving the substrate, are available in the Matlab workspace. Recall that the stochastic rate constant c has to be computed from the deterministic rate constant k according to the relation (5.13). Since, in this

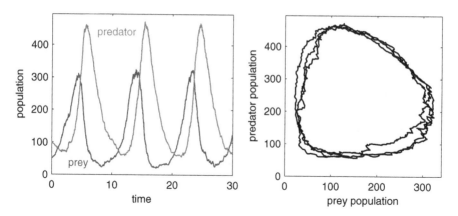

Figure 5.6 Stochastic simulation of the Lotka–Volterra model obtained by one SSA run. *Left*: time course, *Right*: phase plot. Parameters (in sec^{-1}): $\hat{k}_1 = 1$, $\hat{k}_2 = 0.005$, $\hat{k}_3 = 0.6$. Initial populations are taken as 50 individuals of prey and 100 individuals of predator.

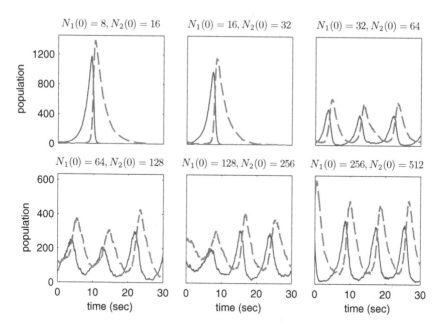

Figure 5.7 Stochastic trajectories of the Lotka–Volterra model of interacting species (2.7) for different initial species populations. The prey and predator populations are plotted in solid and dashed lines, respectively. Note how extinction quickly occurs for some initial populations. Parameters are the same as in Figure 5.6.

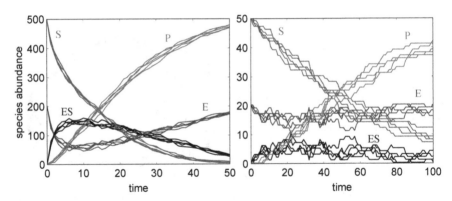

Figure 5.8 Five sample trajectories of species abundance in an enzymatic reaction
(2.8). The volume is chosen as $V = 1.66\,\mathrm{fL}$, so that $\Omega = 1\,(\mathrm{nM})^{-1}$, and hence the
species concentration is numerically the same as the corresponding copy number.
Parameters are taken from Figure 2.7. *Left*: large copy numbers, $n_S^{\mathrm{tot}} = 500$,
$n_E^{\mathrm{tot}} = 200$. *Right*: small copy numbers, $n_S^{\mathrm{tot}} = 50$, $n_E^{\mathrm{tot}} = 20$.

example, only the first reaction channel is bimolecular, we have

$$c_1 = \frac{k_1}{\Omega}, \quad c_2 = k_2, \quad c_3 = k_3,$$

with the Matlab representation

```
ssz = NA*V; % system size
c = [k(1)/ssz, k(2), k(3)];
```

which understands that values of the variables V and NA respectively corresponding to the volume V and Avogadro's number N_A are available in
the Matlab workspace. The volume is chosen to be $V = 1.66\,\mathrm{fL}$, so that
$\Omega = 1\,(\mathrm{nM})^{-1}$, which means that one nanomolar of each species corresponds
to one molecule. To see the variability among realizations, five different
sample trajectories are shown in Figure 5.8 for two scenarios: small/large
initial populations on the left/right. The mean species abundance, together
with the error bars for mean±SD, computed over an ensemble of 10000 realizations, are plotted side by side with the specieswise endpoint empirical
distribution (PMF) in Figure 5.9. Note that the distributions for enzyme
and enzyme–substrate complex have exactly the same shapes and differ only
in their means. This is not a coincidence, but a direct consequence of the
conservation relation (2.24).

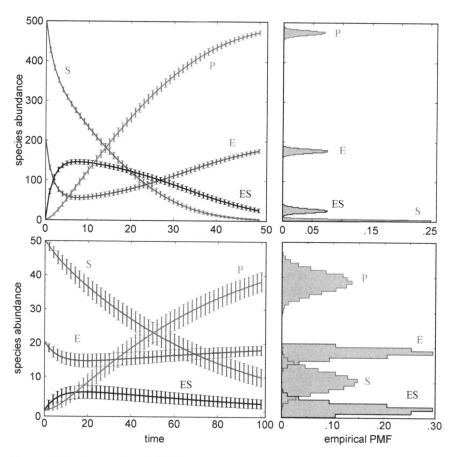

Figure 5.9 Ensemble of 10000 stochastic simulations for the enzymatic reaction (2.8). The parameters and initial copy numbers are taken from Figure 5.8. *Left*: The mean species abundance, together with the error bars according to mean±SD. *Right*: Specieswise endpoint empirical distribution (PMF). *Top*: large copy numbers, $n_S^{\text{tot}} = 500$, $n_E^{\text{tot}} = 200$. *Bottom*: small copy numbers, $n_S^{\text{tot}} = 50$, $n_E^{\text{tot}} = 20$.

Example 5.11 (Stochastic focusing) The branched reaction network (2.10):

$$\varnothing \underset{k_d}{\overset{k_s}{\rightleftharpoons}} S, \quad \varnothing \underset{k_a X_S}{\overset{k_i}{\rightleftharpoons}} I \xrightarrow{k_p} P \xrightarrow{1} \varnothing,$$

contains a two-reaction module

$$\varnothing \xleftarrow{k_a X_S} I \xrightarrow{k_p} P,$$

where intermediate I-molecules have two destination states in which they can end up: state P or state $\emptyset$. For convenience, the system size is taken as $\Omega = 1$. If the two reactions are fast enough, the pool of I-molecules is insignificant, and X_S does not change significantly during the life span of an individual I-molecule, then we can assume the steady state of ending up in P or state $\emptyset$ to be reached immediately. In the steady state, one of the two transitions has occurred, and the probability that it ended up in state P is simply

$$P_P^{ss} = \frac{k_p}{k_p + k_a X_S} = \frac{1}{1 + X_S/K},$$

where $K = k_p/k_a$ is the inhibition constant. A more rigorous derivation of the above result will appear in Section 5.5. The two-reaction module can then be approximated by a single fast reaction $I \rightarrow P$ with transition probability $1/(1+X_S/K)$ taken as the stationary probability of ending up in state P from the last example. Note that we have assigned to the fast reaction an effective transition probability and not a propensity, because the stationary state is assumed to be achieved fast. The simplified transition $I \rightarrow P$ follows in series the transition $\emptyset \xrightarrow{k_i} I$. The two in-series transitions can be combined into one overall transition from $\emptyset$ to I with an effective transition rate $k_i/(1+X_S/K)$. The branched reaction scheme (2.10) then simplifies to (2.29):

$$\emptyset \underset{k_d}{\overset{k_s}{\rightleftharpoons}} S, \quad \emptyset \underset{1}{\overset{k_i/(1+X_S/K)}{\rightleftharpoons}} P.$$

For this simplified scheme, the reaction propensities $a(n)$ in state $n = (n_S, n_P)^T$ are listed here (on the right) together with the corresponding reactions (on the left):

$$
\left.
\begin{array}{ll}
\emptyset \xrightarrow{k_s} S, & a_s^+(n) = k_s, \\[2mm]
S \xrightarrow{k_d} \emptyset, & a_s^-(n) = k_d n_S, \\[2mm]
\emptyset \xrightarrow{k/(1+X_S/K)} P, & a_p^+(n) = \dfrac{k_i}{1 + \frac{n_S}{K}}, \\[2mm]
P \xrightarrow{1} \emptyset, & a_p^-(n) = n_P.
\end{array}
\right\}
\tag{5.24}
$$

The results of stochastic simulations performed for noise-free and noisy signal concentrations are shown in Figure 5.10. When signal noise is insignificant, a twofold decrease in the average signal can never result in more than a twofold increase in the average product abundance due to intrinsic limitations of hyperbolic inhibition. However, when signal noise is significant, a twofold decrease in the average signal concentration results in more than a threefold increase

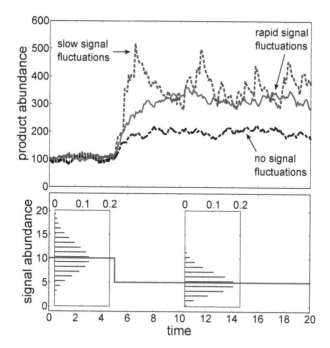

Figure 5.10 Stochastic focusing—fluctuation enhanced sensitivity. Number of product molecules evolving against time when the product formation is inhibited by a noisy or noise-free signal. After five time units, the signal mean $\langle N_S \rangle$ shifts from 10 to 5 due to a twofold reduction in k_s from $10k_d$ to $5k_d$. The respective values of k_d for slow and fast fluctuations are taken as 100 and 1000. Rapid signal fluctuations correspond to insignificant time correlations, and hence negligible product fluctuations. Slow signal fluctuations result in considerable time correlations in the product synthesis rate, and hence significant product fluctuations. Parameters that stay the same before/after the shift are $k_i = 10^4$ and $K = 0.1$. Figure adapted from [121].

increase in the average product abundance (Figure 5.10, *top*). Consequently, when the stationary signal abundance distributions overlap significantly (Figure 5.10, *bottom*), the corresponding average reaction probabilities can in fact become more separated than when fluctuations are negligible: *stochastic focusing*. The increased capacity for sensitivity amplification is the only effect of a rapidly fluctuating signal. Rapid signal fluctuations correspond to insignificant time correlations, and hence negligible product fluctuations. Slow signal fluctuations result in considerable time correlations in the product synthesis rate, and hence significant product fluctuations.

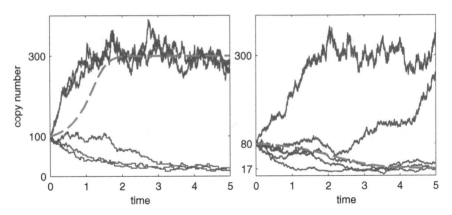

Figure 5.11 Stochastic simulation of the Schlögl model (2.9). Five sample trajectories together with the deterministic time course (dashed). While the deterministic time course settles into one of the two stable fixed points, some of the SSA trajectories spread out to other states. *Left*: Initial copy number $N(0) = 100$ is in the basin of attraction of the first stable fixed point $n = 300$. *Right*: Initial copy number $N(0) = 80$ is in the basin of attraction of the second stable fixed point $n = 17$.

Example 5.12 (Schlögl model) The Schlögl model (2.9) with propensities in (5.19) is a bistable system with two stable steady states separated by an unstable steady state. In a deterministic framework, such a system settles to the steady state whose basin of attraction is nearer the initial condition. In a stochastic framework, however, the behavior is more complex: either steady state may be reached in different realizations regardless of the initial condition. This behavior, referred to as "stochastic switching" in [58, 160], is illustrated here in Figure 5.11, wherein two sets of five sample trajectories, each set starting from a different initial copy number, are plotted side by side. The associated deterministic time course is overlaid on each set. It is easy to see that, while the deterministic time course settles into one of the stable fixed points, some of the stochastic trajectories spread out to other states. This can be more easily seen in the histogram of Figure 1.6, discussed earlier in the introductory chapter. The time-varying histogram, which was obtained from 10000 realizations, is unimodal initially and has a bimodal pattern at the end. The Matlab implementation of the Schlögl model (2.9) with propensities (5.19) in terms of fields of the R structure is left as an exercise for the reader.

Example 5.13 (Gene regulation) Let us revisit the gene regulation model (2.11) with propensities in (5.20). The analysis presented here was motivated from a more detailed discussion of stochasticity (and its origins) in [74], which is highly recommended for a thorough reading. Results of stochastic

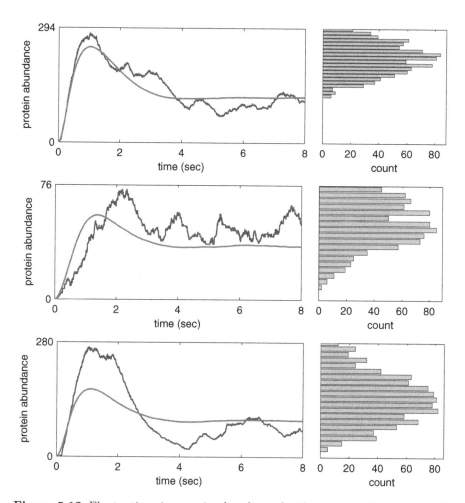

Figure 5.12 Fluctuations in protein abundance for the gene regulatory network (2.11). *Left*: Time courses of a single SSA run (the staircase) and the mean over 1000 runs (the curve). *Right*: Endpoint histogram showing the empirical probability that a cell will have a given protein abundance. The initial conditions are ten copies of the gene G and zero copies of the other species. The rate parameters are $k_m = k_p = 20\,\mathrm{sec}^{-1}$, $k_b = 0.2\,\mathrm{sec}^{-1}$, $k_u = k_p^- = 1\,\mathrm{sec}^{-1}$, and $k_m^- = 1.5\,\mathrm{sec}^{-1}$. *Top*: Small fluctuations with high copy numbers of expressed mRNA and protein. *Middle*: Large fluctuations in protein abundance arise from a decrease in the expressed mRNA abundance, and an associated decrease in protein abundance, due to a tenfold decrease in the transcription rate k_m. *Bottom*: Large fluctuations in protein abundance arise from a decrease in the expressed mRNA abundance due to a tenfold decrease in the transcription rate k_m. The translation rate k_p was increased fivefold to avoid any significant decrease in the protein abundance. We can clearly see that the fluctuations are still large, because fluctuations in mRNA abundance are a dominant source of noise in gene expression. This illustrates the propagation of noise from transcription to translation.

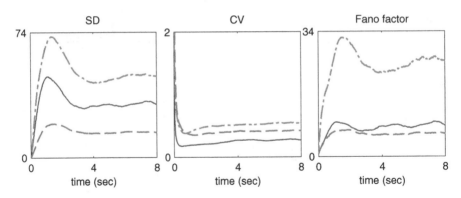

Figure 5.13 Measures of noise in gene regulation model.

simulations are shown in Figure 5.12, where time courses of protein abundance are plotted side by side with the associated empirical probability distributions for three different cases. For the parameters chosen in the first case (Figure 5.12, top), high abundance of expressed mRNA and protein lead to small fluctuations. A tenfold decrease in the transcription rate k_m, the second case (Figure 5.12, middle), leads to a decrease in the expressed mRNA abundance and an associated decrease in protein abundance. That, in turn, leads to large fluctuations in the protein abundance. The third case (Figure 5.12, bottom), highlights how fluctuations in mRNA abundance at the transcription level are a second important factor contributing to gene-expression noise. Therein, in addition to the tenfold decrease in the rate of transcription, the rate of translation is increased fivefold in order to keep the protein abundance more or less the same as in the first case (Figure 5.12, top). The increased gene-expression noise in spite of large protein abundance is attributable to increased fluctuations in mRNA abundance, causing increased fluctuations in the rate of protein synthesis. In other words, noise is propagated from transcription to translation.

Measures of noise: The common mistake of using the standard deviation (SD) as a measure of noise is illustrated in Figure (5.13). The relative levels in the left subplot in Figure (5.13), corresponding to the three cases in the last example, can be misleading because the SD depends on the mean. The relative fluctuation around the mean, measured as the SD divided by the mean, also known as the coefficient of variation (CV), is the most direct and unambiguous measure of gene-expression noise. Note the changed order of levels in the middle subplot in Figure (5.13). It is sometimes advantageous to use a different measure, the *noise strength*, which is defined by the variance divided by the mean and referred to as the "Fano factor". Since the Fano factor is unity for a Poisson process, it can be interpreted as a measure of how close to Poisson a given process is. The Fano factor is used primarily to reveal trends that would otherwise be obscured by noise attributed to low copy numbers; see [74]. Note that although the relative levels in the right subplot in Figure (5.13) are the same as in the left subplot, the separation between levels has changed.

5.4 Continuous Stochastic Simulation

The discrete stochastic simulation method presented in the last section is an exact method. The various improvements and approximations to the original method are all discrete methods. An alternative is to approximate the jump process $N(t)$ by a continuous process $N^c(t)$. Simulating (i.e., generating sample paths of) such a continuous process should be orders of magnitude faster than the discrete simulation. To set the stage for such an approximation, we introduce two important properties of a stochastic process: *drift* and *diffusion*.

5.4.1 Drift and Diffusion

We recall that the mean and (co)variance can be used as crude characterizations of an otherwise unknown probability distribution. When the average value of a stochastic process changes with time, we say that the process is drifting or has a drift. When the (co)variance, a measure of spread of the distribution, changes with time, we say that the process has diffusion. At this stage, we are not in a position to qualitatively discuss the evolving shape of the probability distribution of the (copy-number) process in question. However, the notion of propensity is powerful enough for us to discuss drift and diffusion locally. The two quantities can be best described locally in terms of the increment $\Delta N(t) = N(t + \Delta t) - N(t)$ of the copy-number process $N(t)$ during the time interval $[t, t + \Delta t]$ supposed to be in state n at time t. The short-time drift can be defined as $\langle \Delta N \rangle_n$, the change of the average value of the process during the interval. The short-time diffusion can be defined as $\langle \delta \Delta N \, \delta \Delta N^T \rangle_n$, the change in the covariance of the process during the interval.

An expression for the short-time diffusion $\langle \Delta N \rangle_n$ can be derived in the following way. The increment ΔN follows from (2.2) to be a linear combination

$$\Delta N(t) = S \, \Delta Z(t) \qquad (5.25)$$

of the short-time reaction-count increments ΔZ_j, whose probability distribution, conditioned on state n, follows from (5.7) to be approximately Poisson with mean (and variance) $a_j(n)\Delta t$ during a sufficiently short interval. In other words, ΔZ is approximately Poisson with mean

$$\langle \Delta Z \rangle_n \approx a(n)\Delta t \tag{5.26}$$

and covariance

$$\left\langle \delta\Delta Z\, \delta\Delta Z^T \right\rangle_n \approx \mathrm{diag}(a(n)\Delta t)\,. \tag{5.27}$$

Here $\mathrm{diag}(a)$, for any vector a, denotes the diagonal matrix with elements of a on the main diagonal. The nondiagonal elements are zero because the reaction channels progress independently of each other during the short interval. Following (5.25) and (5.8), the short-time drift takes the form

$$\langle \Delta N \rangle_n = S\, \langle \Delta Z \rangle_n \approx S\, a(n)\Delta t = A(n)\Delta t, \tag{5.28}$$

where the vector

$$A(n) = S\, a(n) \tag{5.29}$$

is the drift per unit time interval, the *drift rate*,

$$A_i(n) = \sum_{j=1}^{r} S_{ij} a_j(n)\,.$$

The short-time diffusion $\left\langle \delta\Delta N\, \delta\Delta N^T \right\rangle_n$ can be worked out in a similar way as follows:

$$
\begin{aligned}
\left\langle \delta\Delta N\, \delta\Delta N^T \right\rangle_n &= \left\langle \Delta N \Delta N^T \right\rangle_n - \langle \Delta N \rangle \langle \Delta N \rangle^T \\
&= S\left[\left\langle \Delta Z \Delta Z^T \right\rangle_n - \langle \Delta Z \rangle_n \langle \Delta Z \rangle_n^T \right] S^T \\
&= S\left\langle \delta\Delta Z\, \delta\Delta Z^T \right\rangle_n S^T \\
&\approx S\, \mathrm{diag}(a(n)\Delta t)\, S^T\,.
\end{aligned}
$$

Thus the short-time diffusion takes the form

$$\left\langle \delta\Delta N\, \delta\Delta N^T \right\rangle_n \approx B(n)\Delta t, \tag{5.30}$$

where the matrix

$$B(n) = S\, \mathrm{diag}(a(n))\, S^T \tag{5.31}$$

is the diffusion per unit time interval, the *diffusion rate*, with elements

$$B_{ik}(n) = \sum_{j=1}^{r} S_{ij} S_{kj} a_j(n)\,.$$

Drift/diffusion terminology: In the physics literature, the drift/diffusion rate is usually referred to as the "drift/diffusion coefficient" or "drift/diffusion vector/matrix." We believe that the latter (vector/matrix) terms are misleading because they do not reflect the *per unit time* nature of the associated quantities.

5.4.2 Chemical Langevin Equation

Equations (5.26) and (5.28) give useful expressions for the average reaction-count increment $\langle \Delta Z \rangle_n$ and average copy-number increment $\langle \Delta N \rangle_n$, respectively, conditioned on $N(t) = n$. Since we are aiming for a continuous approximation $N^c(t)$ of the jump process $N(t)$ itself rather than averages, we need expressions for the copy-number increment

$$(\Delta N)_n = N(t + \Delta t) - n$$

from the fixed state $N(t) = n$ and the related reaction-count increment $(\Delta Z)_n$. We pointed out in the last subsection that $(\Delta Z_j)_n$ has a Poisson distribution with mean $a_j(n)\Delta t$ during the short time interval. The requirement of Δt to be small enough to assume a constant propensity during the interval was reported as condition (i) in [56]. The property of the Poisson distribution to approach a normal distribution for a very large mean motivates the condition (ii) assumed in [56] on Δt: it is large enough for the average reaction count increment $\langle \Delta Z_j \rangle$ of every channel to be very large, $a_j(n)\Delta t \gg 1$, so that the Poisson variable ΔZ_j can be approximated by a normal random variable with the same mean and variance $a_j(n)\Delta t$ for each reaction channel. Since any normal random variable can be written as a sum of its mean and the standard normal variable (with zero mean and unit variance) multiplied by its standard deviation, we can write the normal approximation as

$$(\Delta Z_j)_n \approx a_j(n)\Delta t + (a_j(n)\Delta t)^{1/2}\, \mathcal{N}_j(t), \tag{5.32}$$

where $\mathcal{N}_j(t)$ is a standard normal (i.e., zero mean and unit variance) process associated with reaction channel R_j. The channelwise processes collected in the r-vector process $\mathcal{N}(t)$ are statistically independent. Inserting the above equation into (5.25) gives the copy-number increments

$$(\Delta N_i)_n \approx \sum_{j=1}^{r} S_{ij} a_j(n)\Delta t + \sum_{j=1}^{r} S_{ij}\, (a_j(n)\Delta t)^{1/2}\, \mathcal{N}_j(t).$$

The increment vector then takes the form

$$(\Delta N)_n \approx A(n)\Delta t + D(n)\sqrt{\Delta t}\mathcal{N}(t), \tag{5.33}$$

where we have recognized the drift rate $A(n) = S\,a(n)$ and the newly appearing matrix

$$D(n) = S \operatorname{diag}\left(a(n)\right)^{1/2},\tag{5.34}$$

is referred to, in this text, as the *drift coefficient*. The square root should be interpreted elementwise. The factor $\sqrt{\Delta t}\mathcal{N}(t)$ in the second summation on the right can be recognized as the Wiener increment,

$$\Delta \mathcal{W} = \mathcal{W}(t + \Delta t) - \mathcal{W}(t) = \sqrt{\Delta t}\mathcal{N}(t),$$

of an r-vector $\mathcal{W}(t)$ of independent standard *Brownian motions*, or standard *Wiener processes*. Setting $\Delta t = dt$ followed by replacing $\sqrt{\Delta t}\mathcal{N}(t)$ by the Wiener increment $\Delta \mathcal{W}$ and the supposedly fixed n by the continuous approximation $N^c(t)$ of the original jump process $N(t)$, we arrive at an Ito stochastic differential equation (SDE),

$$dN^c(t) = A\left(N^c(t)\right)dt + D\left(N^c(t)\right)d\mathcal{W}(t).\tag{5.35}$$

In the econometrics literature, the coefficients $A(n)$ and $D(n)$ are respectively referred to as the (vector-valued) "drift-rate function" and the (matrix-valued) "diffusion-rate function." This is a bit unfortunate, because diffusion rate refers, in the physical sciences, to the matrix $B(n)$ in line with the present text. That is why we have adopted the term "drift coefficient" for $D(n)$. With its coefficients specifically defined by $A(n) = S\,a(n)$ and (5.34), the SDE (5.35) is known as the *chemical Langevin equation* (CLE) in the "standard form." An equivalent "white-noise form" of CLE is

$$\frac{dN^c(t)}{dt} = A\left(N^c(t)\right) + D\left(N^c(t)\right)\Gamma(t),\tag{5.36}$$

where the elements of $\Gamma(t) = d\mathcal{W}/dt$ are statistically independent Gaussian white-noise processes.

In the above (two forms of) CLE, we have one Wiener process $\mathcal{W}_j$ for each reaction channel. An alternative form of the CLE can be derived that has a Wiener process $\mathcal{W}_i$ for each chemical component. The derivation goes like this. Being a linear combination of Gaussian process $(\Delta Z)_n$, the copy-number increment $(\Delta N)_n$ is also a Gaussian variable of the form

$$(\Delta N)_n \approx \langle \Delta N \rangle_n + \left\langle \delta \Delta N\, \delta \Delta N^T \right\rangle_n^{1/2} \mathcal{N}(t),\tag{5.37}$$

which is a sum of the mean and an $s \times 1$ standard Gaussian random vector $\mathcal{N}(t)$ premultiplied by the matrix square root of the covariance matrix. Here the matrix square root $M^{1/2}$ of a matrix M is defined such that $M = M^{1/2}\left(M^{1/2}\right)^T$. Inserting expressions (5.28) and (5.30) into the above equation gives

$$(\Delta N)_n \approx A(n)\Delta t + (B(n))^{1/2}\,\mathcal{N}(t)\,(\Delta t)^{1/2}\ . \tag{5.38}$$

Replacing the supposedly known n by $N^c(t)$ and recognizing the Wiener increments $\Delta W = \mathcal{N}(t)\sqrt{\Delta t}$, followed by substitution $\Delta t = \mathrm{d}t$, we obtain the alternative form

$$\mathrm{d}N^c = A\,(N^c)\,\mathrm{d}t + \left(B\,(N^c)\right)^{1/2}\,\mathrm{d}W \tag{5.39}$$

of the CLE. The matrix square root $B^{1/2}$ can be computed from the eigenvalue decomposition of the diffusion rate B.

   The two conditions (i) and (ii) seem conflicting and require the existence of a domain of macroscopically infinitesimal time intervals. Although the existence of a such a domain cannot be guaranteed, Gillespie argues that this can be found for most practical cases. Admitting that, "it may not be easy to continually monitor the system to ensure that conditions (i) and (ii) [...] are satisfied." He justifies his argument by saying that this "will not be the first time that Nature has proved to be unaccommodating to our purposes" [56].

   Generating sample paths of (5.36) is orders of magnitude faster than doing the same for the CME, because it essentially needs generation of normal random numbers. See [68] for numerical simulation methods of stochastic differential equations such as (5.35) and (5.39). The choice between (5.35) and (5.39) may be dictated by the number of reactions r relative to the number of species s, because in each simulation step, the former requires r random numbers, whereas the latter requires s random numbers.

Matlab implementation: Since a CLE is just a special form of the more general SDE, we can use the well-known "Euler–Maruyama method" found in, for example, [68] for generating sample paths of the stochastic process represented by a CLE. The method involves generation of random numbers from the Gaussian distribution to represent the Wiener processes and then using the update rule (5.35) at each time step. The Euler–Maruyama method is implemented here in M-code 5.3. The reader is recommended to read [68] for more efficient methods and implementations.

Example 5.14 (Standard modification) For the standard modification (2.5) with propensities in (5.14), the drift rate is the scalar

$$A(n) = -a_w + a_u = -k_w n + (n^{\mathrm{tot}} - n)k_u,$$

M-code 5.3 simCLE: computes CLE sample paths by the Euler–Maruyama method. Requires the statistics toolbox.

```
function [T,N] = simCLE(S,a,n0,tmax,dt)
[s,r] = size(S);
% Function to generate a sample path of CLE
A = @(n) S*a(n); % Drift rate
D = @(n) S*diag(sqrt(a(n))); % Diffusion coefficient
Steps = tmax/dt; % number of steps
T = (0:dt:tmax)'; % Time output
% Refresh the random number generator
strm = RandStream('mt19937ar','Seed',142857);
RandStream.setDefaultStream(strm);
dW = sqrt(dt)*randn(r,1); % Wiener processes
n = n0;
N = zeros(Steps+1,s); % time course of states
N(1,:) = n0;
for i=2:Steps
    n = n + dt*A(n) + D(n)*dW;
    N(i,:) = real(n);
end
```

and the diffusion coefficient is the 1×2 matrix

$$D(n) = \begin{bmatrix} -\sqrt{a_w} & \sqrt{a_u} \end{bmatrix} = \begin{bmatrix} -\sqrt{k_w n} & \sqrt{(n^{\text{tot}} - n)k_u} \end{bmatrix}.$$

With $A(n)$ and $D(n)$ available, the CLE (5.35) gives us an update rule to generate the next state from the current state in a simulation. Specifically, when the current state is n, the next state (after one time step dt used in the simulation) will be

$$n + A(n)\mathrm{d}t + D(n)\mathrm{d}\mathcal{W}(t).$$

Approximate trajectories can be generated using the above CLE update rule in any standard simulation environment such as Matlab. Figure 5.14 shows time courses (on the left) plotted side by side with the endpoint histogram (on the right), both computed from CLE simulations and overlaid on the same quantities computed with SSA. It can be seen that the mean±SD of 1000 realizations computed with CLE simulations match closely the same quantities computed with SSA. Although not an exact match, the essence of both the time course and the probability distribution is captured by the CLE.

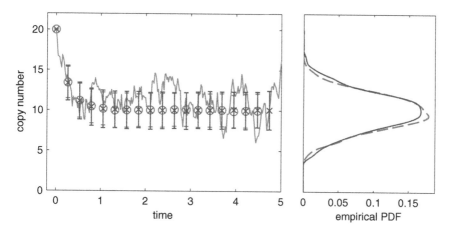

Figure 5.14 CLE simulation for the standard modification (2.5). *Left*: A single CLE run (fluctuating) and mean over 1000 runs together with the error bars determined by mean±SD (threads). The CLE computed mean (*stars*) as well as the error bars match closely the ensemble mean (*circles*), and the associated error bars, that were computed over 1000 SSA runs. *Right*: Endpoint histogram. The empirical probability density computed with CLE simulations (*solid* curve) matches closely the density computed with SSA runs. Parameters and initial conditions are the same as in Figure 5.3.

Example 5.15 (Heterodimerization) For the heterodimerization (2.6) with propensities in (5.15), the drift rate is the scalar

$$A(n) = a_1 - a_2 = \hat{k}_1\left(\hat{q}_1 - n\right)\left(\hat{q}_2 - n\right) - k_2 n,$$

and the diffusion coefficient is the 1×2 matrix

$$D(n) = \left[\begin{array}{cc} \sqrt{a_1} & -\sqrt{a_2} \end{array}\right] = \left[\begin{array}{cc} \sqrt{\hat{k}_1\left(\hat{q}_1 - n\right)\left(\hat{q}_2 - n\right)} & -\sqrt{k_2 n} \end{array}\right].$$

With $A(n)$ and $D(n)$ determined, we can generate approximate trajectories from the CLE (5.35). Figure 5.15 shows time courses (on the left) plotted side by side with the endpoint histogram (on the right), both computed from CLE simulations and overlaid on the same quantities computed with SSA. It can be seen that the mean±SD of 1000 realizations computed with CLE simulations match closely with the same quantities computed with SSA.

Example 5.16 (Lotka–Volterra model) For the Lotka–Volterra model (2.7)

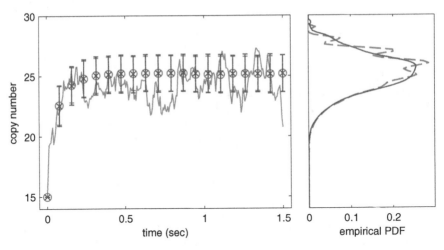

Figure 5.15 CLE simulation for the heterodimerization (2.6). *Left*: A single CLE run (fluctuating) and mean over 1000 runs together with the error bars determined by mean±SD (threads). The CLE computed mean (*stars*) as well as the error bars match closely the ensemble mean (*circles*), and the associated error bars, that were computed over 1000 SSA runs. *Right*: Endpoint histogram. The empirical probability density computed with CLE simulations (*solid* curve) matches closely the density computed with SSA runs. Parameters and initial conditions are the same as in Figure 5.5.

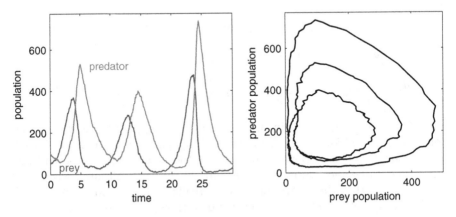

Figure 5.16 CLE simulation of the Lotka–Volterra model obtained by one CLE run. *Left*: time course, *Right*: phase plot. Parameters and initial conditions are the same as in Figure 5.6.

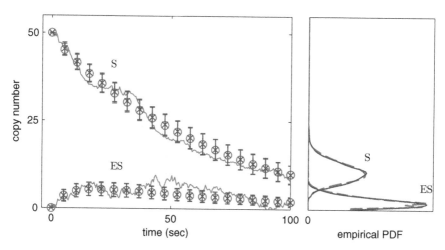

Figure 5.17 CLE simulation for the enzymatic reaction (2.8). *Left*: A single CLE run (fluctuating) and mean over 1000 runs together with the error bars determined by mean±SD (threads). The CLE computed mean (*stars*) as well as the error bars match closely the ensemble mean (*circles*), and the associated error bars, that were computed over 1000 SSA runs. *Right*: Endpoint histogram. The empirical probability density computed with CLE simulations (*solid* curve) matches closely with the density computed SSA runs. Parameters and initial conditions are the same as in Figure 2.7.

with propensities in (5.16), the drift rate is the 2×1 vector

$$
A(n) = \begin{bmatrix} a_1 - a_2 \\ a_2 - a_3 \end{bmatrix} = \begin{bmatrix} \hat{k}_1 n_A n_1 - \hat{k}_2 n_1 n_2 \\ \hat{k}_2 n_1 n_2 - \hat{k}_3 n_2 \end{bmatrix},
$$

and the diffusion coefficient is the 2×2 matrix

$$
D(n) = \begin{bmatrix} \sqrt{a_1} & -\sqrt{a_2} \\ \sqrt{a_2} & -\sqrt{a_3} \end{bmatrix} = \begin{bmatrix} \sqrt{\hat{k}_1 n_A n_1} & -\sqrt{\hat{k}_2 n_1 n_2} \\ \sqrt{\hat{k}_2 n_1 n_2} & -\sqrt{\hat{k}_3 n_2} \end{bmatrix}.
$$

After determining A and D, we can generate approximate trajectories from the CLE (5.35). Figure 5.16 shows time courses (on the left) and endpoint histogram (on the right), both CLE generated. This figure should be compared with Figure 5.6, which shows SSA-computed plots.

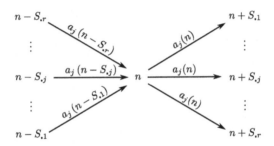

Figure 5.18 State transitions of a generic r-reaction network with network structure encoded in the stoichiometry matrix S and reaction kinetics encoded in the propensity function $a(n)$.

Example 5.17 (Enzyme kinetic reaction) For the enzymatic reaction (2.8) with propensities in (5.17), the drift rate is the 2×1 vector

$$
A(n) = \begin{bmatrix} a_2 - a_1 \\ a_1 - a_2 - a_3 \end{bmatrix} = \begin{bmatrix} k_2 n_{\mathrm{ES}} - \hat{k}_1 \left(n_{\mathrm{E}}^{\mathrm{tot}} - n_{\mathrm{ES}} \right) n_{\mathrm{S}} \\ \hat{k}_1 \left(n_{\mathrm{E}}^{\mathrm{tot}} - n_{\mathrm{ES}} \right) n_{\mathrm{S}} - (k_2 + k_3) n_{\mathrm{ES}} \end{bmatrix},
$$

and the diffusion coefficient is the 2×2 matrix

$$
\begin{aligned}
D(n) &= \begin{bmatrix} -\sqrt{a_1} & \sqrt{a_2} & \sqrt{a_3} \\ \sqrt{a_1} & -\sqrt{a_2} & -\sqrt{a_3} \end{bmatrix} \\
&= \begin{bmatrix} -\sqrt{\hat{k}_1 \left(n_{\mathrm{E}}^{\mathrm{tot}} - n_{\mathrm{ES}} \right) n_{\mathrm{S}}} & \sqrt{k_2 n_{\mathrm{ES}}} & \sqrt{k_3 n_{\mathrm{ES}}} \\ \sqrt{\hat{k}_1 \left(n_{\mathrm{E}}^{\mathrm{tot}} - n_{\mathrm{ES}} \right) n_{\mathrm{S}}} & -\sqrt{k_2 n_{\mathrm{ES}}} & -\sqrt{k_3 n_{\mathrm{ES}}} \end{bmatrix}.
\end{aligned}
$$

Once A and D are available, we can generate approximate trajectories from the CLE (5.35). Figure 5.17 shows time courses (on the left) plotted side by side with the endpoint histogram (on the right), both computed from CLE simulations and overlaid on the same quantities computed with SSA. It can be seen that the mean±SD of 1000 realizations computed with CLE simulations match closely the same quantities computed with SSA.

5.5 Chemical Master Equation

The occurrence of each reaction moves the system from one state to another in the state space. The possible state transitions from/to state n are usually sketched in a state transition diagram like the one in Figure 5.18, where a transition from one state to another is represented by an arrow labeled with the corresponding transition rate. The transition rate of a state transition resulting from a single reaction channel is equal to the reaction propensity of that channel. The transition rate of a state transition resulting from more than one reaction channel is the sum of propensities of those reaction channels.

How does the state probability $P(n,t)$ change with time? To answer this, we need to find an expression for $P(n, t + \Delta t)$, the probability of being in state n after a short time interval of length Δt. How can the system land in state n at time $t + \Delta t$? One possibility is that the system was in state n at time t and no reaction occurred during the interval. Otherwise, as obvious from the state transition diagram in Figure 5.18, the state n was reached after the occurrence of one of r possible reactions. Mathematically, we can write

$$P(n|n', \Delta t) = o(\Delta t) + \begin{cases} 1 - a_0(n)\Delta t & \text{if } n' = n, \\ a_1\left(n - S_{.1}\right)\Delta t & \text{if } n' = n - S_{.1}, \\ \vdots & \\ a_r\left(n - S_{.r}\right)\Delta t & \text{if } n' = n - S_{.r}, \\ 0 & \text{otherwise.} \end{cases}$$

The term $o(\Delta t)$ represents the probability of arriving in state n by the occurrence of more than one reaction during the interval. Recall that $a_0(n) = \sum_j a_j(n)$ is the exit rate from state n. Substituting the above expressions into (5.3) gives

$$P(n, t + \Delta t) = o(\Delta t) + P(n,t)\left(1 - \sum_{j=1}^{r} a_j(n)\Delta t\right)$$

$$+ \sum_{j=1}^{r} P\left(n - S_{.j}, t\right) a_j\left(n - S_{.j}\right)\Delta t,$$

which for vanishingly short Δt can be rearranged as the *chemical master equation* (CME):

$$\frac{\partial}{\partial t}P(n, t) = \sum_{j=1}^{r}\left[a_j\left(n - S_{.j}\right)P\left(n - S_{.j}, t\right) - a_j(n)P(n, t)\right]. \qquad (5.40)$$

$$n-1 \; \underset{k_w n}{\overset{k_u \left(n^{\text{tot}} - n + 1\right)}{\rightleftarrows}} \; n \; \underset{k_w(n+1)}{\overset{k_u \left(n^{\text{tot}} - n\right)}{\rightleftarrows}} \; n+1$$

Figure 5.19 State transitions of the standard modification (2.5).

We should note two important facets of the way we have written the CME above. First, in spite of the use of a partial differential operator ∂ and the appearance of the CME as a differential–difference equation (*differential* in time t and *difference* in states n), it is a large (typically infinite-dimensional) system of ODEs because n is a sample (or state) of the copy number $N(t)$, and therefore we need an ODE for each state. Since the state n is fixed, a more intuitive notation would place it as a subscript, namely $P_n(t)$, and replace the partial differential operator ∂ with an ordinary differential operator d. That would, however, lead to notational complications in the subsequent sections where we will be extending the state space of n to the nonnegative real line and will require differentiation of probabilities with respect to n. Second, recall that the CME above has been written with an understanding that the functional form of the propensities $a_j(n)$ has been specified for the process under study. Without that specification, the CME, similar to the CKE, merely represents a consistency condition imposed by the Markov property.

5.5.1 The Dimensionality Curse

Since there is potentially a large number of possible states, any attempt to solve the CME analytically or even numerically will be impractical, unless one is dealing with a very simple system such as the isomerization reaction (2.5) that has just one state variable n and only two channels (state transitions). When all the reactions in a system are monomolecular, it is possible to solve the CME analytically, although the procedure is quite involved mathematically [73]. However, bimolecular reactions typically occur more often in biological processes because of enzymatic associations, making the CME intractable for such systems. This means that one has to resort to either stochastic simulations (covered in the previous section), analytical approximations (to come in the following section), or numerical approximations including the "finite state projection algorithm" [103] and the "sliding windows" method [167].

Example 5.18 (Standard modification) For the standard modification (2.5) with propensities in (5.14), the state transition diagram is given in Figure

5.19. Based on these state transitions, the CME for this example reads

$$\frac{\partial}{\partial t} P(n,t) = k_w \left[(n+1)P(n+1,t) - nP(n,t) \right]$$
$$+ k_u \left[(n^{\text{tot}} - n + 1) P(n-1,t) - (n^{\text{tot}} - n) P(n,t) \right]. \quad (5.41)$$

Note that this CME must respect the boundary conditions with respect to $n = 0, 1, \ldots, n^{\text{tot}}$. That is, $P(n,t) = 0$ for $0 > n > n^{\text{tot}}$. We can gain some insight into the dynamics described in the above CME by setting $n^{\text{tot}} = 1$, which corresponds to a single molecule (in isolation) that can exist either in the unmodified form U with probability $P_U(t) \stackrel{\text{def}}{=} P(1,t)$, or in the modified form W with probability $P_W(t) \stackrel{\text{def}}{=} P(0,t) = 1 - P_U(t)$. The single-molecule version of the above CME turns out to be

$$\frac{\mathrm{d}}{\mathrm{d}t} P_U(t) = -k_w P_U(t) + k_u (1 - P_U(t)) = k_u - (k_w + k_u) P_U(t),$$

where we have used the boundary condition $P(2,t) = 0$. Suppose that the protein molecule is initially unmodified, that is, $P_U(0) = 1$. Then the above single molecule CME can be solved for $P_U(t)$ to yield

$$P_U(t) = \frac{k_u + k_w e^{-(k_w + k_u)t}}{k_w + k_u}.$$

Having determined the probability P_U of a single molecule to be unmodified, the probability $P(n,t)$ that n out of all the available n^{tot} are unmodified is simply the PMF of the binomial distribution $\text{Binomial}(n^{\text{tot}}, P_U)$, namely

$$P(n,t) = \binom{n^{\text{tot}}}{n} (P_U(t))^n (1 - P_U(t))^{n^{\text{tot}} - n}.$$

We have thus found the solution to the original CME (5.41) through an indirect, but insightful, procedure. This will, however, not be tractable for every case. The progress, in time, of the probability distribution of the copy number $N(t)$ (of molecules in inactive form) is shown in Figure 5.20, wherein the PMF is plotted during two time subintervals.

Example 5.19 (Hyperbolic control) Hyperbolic control arises from a multitude of schemes, for instance

$$\text{P} \xleftarrow{\ k_p\ } \text{I} \xrightarrow{\ k_a X_S\ } \text{A}.$$

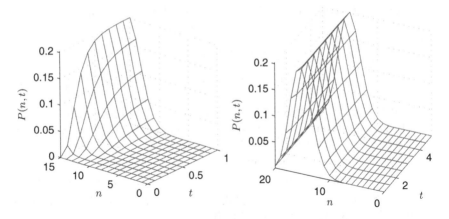

Figure 5.20 Temporal progress of the probability distribution for the standard modification. The PMF $P(n,t)$, for the copy number of unmodified proteins, is plotted during two time subintervals: $0 < t < 1$ (*left*) and $1 < t < 5$ (*right*). The parameters were chosen as $k_w = 3$ and $k_u = 1$, both in $\sec^{-1}$. Initially, all the proteins are assumed to be unmodified.

Here the usual conversion, with rate coefficient k_p, of I-molecule to P-molecule (the product) is inhibited by a signaling S-molecule that binds to an I-molecule and transforms it to an A-molecule with a rate coefficient $k_a X_S$ that depends on signal concentration X_S. Suppose that $X_S(t) = x_S$ at time point t. At this time point, the I-molecule is either unchanged from its original I-form with probability $P_I(t)$, has converted to the product with probability $P_P(t)$, or has been converted to an A-molecule with probability

$$P_A(t) = 1 - P_I(t) + P_P(t) \,.$$

It is important to keep in mind that the probabilities here are conditioned on $X_S(t) = x_S$. The master equations for $P_P(t)$ and $P_I(t)$ can be written as

$$\dot{P}_P(t) = k_p P_I(t), \quad \dot{P}_I(t) = -(k_p + k_a x_S) P_I(t) \,.$$

The solution of the second master equation, subject to the obvious initial condition $P_I(0) = 1$, is

$$P_I(t) = \exp\left(-(k_p + k_a x_S) t\right) \,.$$

Substituting this solution into the first master equation and integrating the latter during time interval $[0, t]$ yields its solution:

$$P_P(t) = \frac{k_p}{k_p + k_a x_S} \left[1 - \exp\left(-\left(k_p + k_a x_S\right)t\right)\right].$$

The probability $P_I(t)$ will eventually vanish, leaving the stationary probability distribution

$$P_P^{ss} = \frac{k_p}{k_p + k_a x_S} = \frac{1}{1 + x_S/K}, \quad P_A^{ss} = \frac{x_S/K}{1 + x_S/K},$$

the probabilities of ending up in state P and A, respectively. The new parameter $K = k_p/k_a$ can be interpreted as an inhibition constant. As noted in the previous example, the above steady-state probabilities could have been obtained by intuitive reasoning under the assumption of chemical equilibrium: the respective fractions of time spent in states P and A are $k_p/(k_p+k_a x_S)$ and $k_a x_S/(k_p+k_a x_S)$.

Example 5.20 (Stochastic focusing) The two-reaction module in the last example appears in the branched reaction network (2.10):

$$\emptyset \underset{k_d}{\overset{k_s}{\rightleftharpoons}} S, \quad \emptyset \underset{k_a X_S}{\overset{k}{\rightleftharpoons}} I \xrightarrow{k_p} P \xrightarrow{1} \emptyset,$$

where the A-molecules appear as the null species $\emptyset$. When the transition from I to $\emptyset$ or P is fast enough, the pool of I-molecules is insignificant, and X_S does not change significantly during the life span of an individual I-molecule. The two-reaction module

$$\emptyset \xleftarrow{k_a X_S} I \xrightarrow{k_p} P$$

can then be approximated by a single fast reaction $I \to P$ with transition probability $1/(1+x_S/K)$ taken as the stationary probability of ending up in state P from the last example. Note that we have assigned to the fast reaction an effective transition probability and not a propensity, because the stationary state is assumed to be achieved fast. The simplified transition $I \to P$ follows in series the transition $\emptyset \xrightarrow{k_i} I$. The two in-series transitions can be combined into one overall transition from $\emptyset$ to I with an effective transition rate $k_i/(1+x_S/K)$. The branched reaction scheme (2.10) then simplifies to (2.29):

$$\emptyset \underset{k_d}{\overset{k_s}{\rightleftharpoons}} S, \quad \emptyset \underset{1}{\overset{k_i/(1+X_S/K)}{\rightleftharpoons}} P,$$

with the state transition diagram shown in Figure 5.21. Based on these state

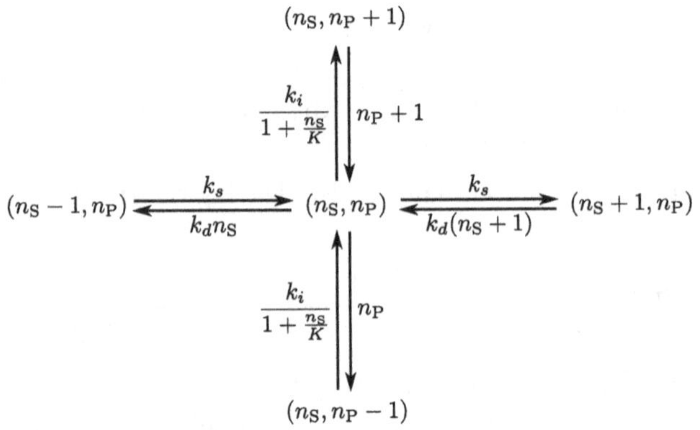

$$(n_S, n_P + 1)$$

$$\dfrac{k_i}{1 + \frac{n_S}{K}} \Big\| n_P + 1$$

$$(n_S - 1, n_P) \xrightleftharpoons[k_d n_S]{k_s} (n_S, n_P) \xrightleftharpoons[k_d(n_S + 1)]{k_s} (n_S + 1, n_P)$$

$$\dfrac{k_i}{1 + \frac{n_S}{K}} \Big\| n_P$$

$$(n_S, n_P - 1)$$

Figure 5.21 State transitions of the simplified branched network (2.29) (stochastic focusing).

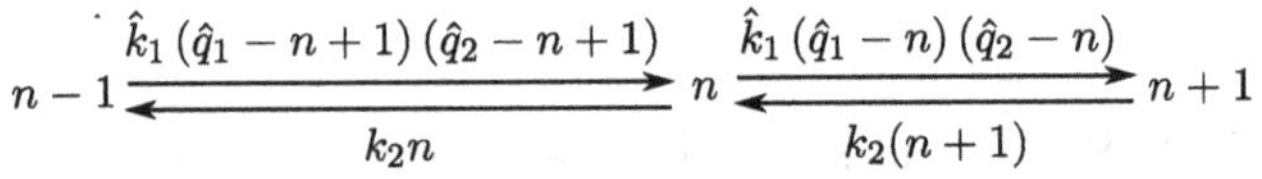

$$n - 1 \xrightleftharpoons[k_2 n]{\hat{k}_1 (\hat{q}_1 - n + 1)(\hat{q}_2 - n + 1)} n \xrightleftharpoons[k_2(n+1)]{\hat{k}_1 (\hat{q}_1 - n)(\hat{q}_2 - n)} n + 1$$

Figure 5.22 State transitions of the heterodimerization reaction (2.6).

transitions, the CME for the simplified scheme can be derived to read

$$\frac{\partial}{\partial t} P(n_S, n_P, t) = k_s P(n_S - 1, n_P, t) - k_s P(n_S, n_P, t)$$
$$+ k_d (n_S + 1) P(n_S + 1, n_P, t) - k_d n_S P(n_S, n_P, t)$$
$$+ \frac{k_i}{1 + n_S/K} \Big[P(n_S, n_P - 1, t) - P(n_S, n_P, t) \Big]$$
$$+ (n_P + 1) P(n_S, n_P + 1, t) - n_P P(n_S, n_P, t) .$$

Example 5.21 (Heterodimerization) For the reversible heterodimerization (2.6) with propensities (5.15), the state transition diagram is shown in Figure 5.22. Based on these state transitions, the CME for this example reads

$$\frac{\partial}{\partial t} P(n, t) = \hat{k}_1 (\hat{q}_1 - n + 1)(\hat{q}_2 - n + 1) P(n - 1, t)$$
$$- \hat{k}_1 (\hat{q}_1 - n)(\hat{q}_2 - n) P(n, t)$$
$$+ k_2 \Big[(n + 1) P(n + 1, t) - n P(n, t) \Big],$$

where $\hat{q}_1 = \Omega q_1$ and $\hat{q}_2 = \Omega q_2$ are the conserved copy numbers, and $\Omega = N_A V$

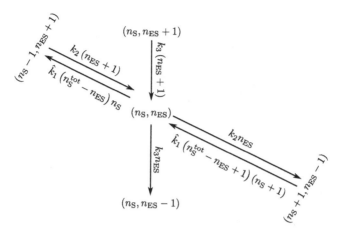

Figure 5.23 State transitions of the enzyme kinetic reaction (2.8).

is the system size..

Example 5.22 (Enzyme kinetic reaction) For the enzymatic reaction (2.8) with propensities in (5.17), the state transition diagram is given in Figure 5.23. Following these state transitions, the CME can be written as

$$\frac{\partial}{\partial t} P\left(n_S, n_{ES}, t\right) = \hat{k}_1 \left(n_S^{tot} - n_{ES} + 1\right)\left(n_S + 1\right) P\left(n_S + 1, n_{ES} - 1, t\right)$$
$$- \hat{k}_1 \left(n_S^{tot} - n_{ES}\right) n_S P\left(n_S, n_{ES}, t\right)$$
$$+ k_2 \Big[\left(n_{ES} + 1\right) P\left(n_S - 1, n_{ES} + 1, t\right) - n_{ES} P\left(n_S, n_{ES}, t\right)\Big]$$
$$+ k_3 \Big[\left(n_{ES} + 1\right) P\left(n_S, n_{ES} + 1, t\right) - n_{ES} P\left(n_S, n_{ES}, t\right)\Big].$$

Example 5.23 (Lotka–Volterra model) For the Lotka–Volterra model (2.7) with propensities in (5.16), the state transition diagram is given in Figure 5.24. From these state transitions, the CME for this example reads

$$\frac{\partial}{\partial t} P\left(n_1, n_2, t\right) = k_1 \Big[\left(n_1 - 1\right) P\left(n_1 - 1, n_2, t\right) - n_1 P\left(n_1, n_2, t\right)\Big]$$
$$+ k_2 \Big[\left(n_1 + 1\right)\left(n_2 - 1\right) P\left(n_1 + 1, n_2 - 1, t\right) - n_1 n_2 P\left(n_1, n_2, t\right)\Big]$$
$$+ k_3 \Big[\left(n_2 + 1\right) P\left(n_1, n_2 + 1, t\right) - n_2 P\left(n_1, n_2, t\right)\Big].$$

Example 5.24 (Schlögl model) For the Schlögl model (2.9) with propensities in (5.18), the state transition diagram is given in Figure 5.25. But recall

(n1 − 1, n2 + 1)
(n1, n2 + 1)
k2n1n2
k3(n2 + 1)
(n1 − 1, n2)
k1nA(n1 − 1)
(n1, n2)
k1nAn1
(n1 + 1, n2)
k2(n1 + 1)(n2 − 1)
k3n2
(n1, n2 − 1)
(n1 + 1, n2 − 1)

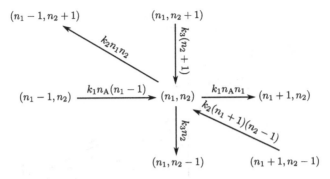

Figure 5.24 State transitions of the Lotka–Volterra model (2.7).

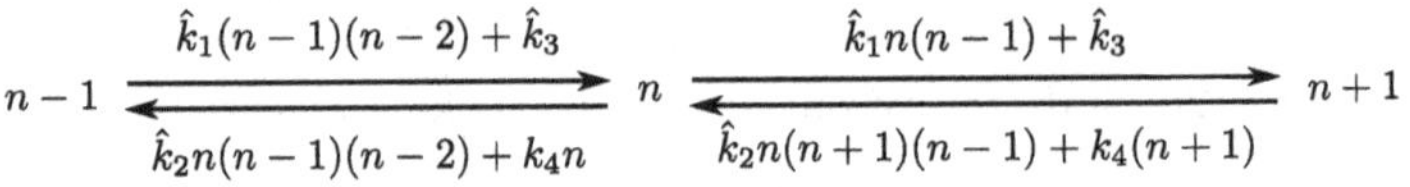

k̂1(n − 1)(n − 2) + k̂3
k̂1n(n − 1) + k̂3
n − 1
n
n + 1
k̂2n(n − 1)(n − 2) + k4n
k̂2n(n + 1)(n − 1) + k4(n + 1)

Figure 5.25 State transitions of the Schlögl model (2.9).

that the same transition diagram also corresponds to the reduced reaction network (5.19). Following these state transitions, the CMEs for both the original Schlögl reaction (5.18) and the reduced reaction network (5.19) read the same:

$$\begin{aligned}\frac{\partial}{\partial t}P(n,t) &= \Big[\hat{k}_1(n-2)(n-1)+\hat{k}_3\Big]P(n-1,t) - \Big[\hat{k}_1(n-1)n+\hat{k}_3\Big]P(n,t) \\ &\quad + \hat{k}_2\Big[(n+1)P(n+1,t)-(n-2)P(n,t)\Big](n-1)n \\ &\quad + k_4\Big[(n+1)P(n+1,t)-nP(n,t)\Big]\,.\end{aligned}$$

Example 5.25 (Gene regulation) For the gene regulation scheme (2.11) with propensities in (5.20), the state transition diagram is given in Figure 5.26.

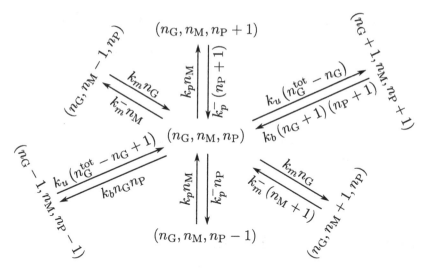

Figure 5.26 State transitions of the gene regulation (2.11).

Following these state transitions, the CME can be written as

$$\frac{\partial}{\partial t} P(n_G, n_M, n_P, t)$$

$$= k_m n_G \left[P(n_G, n_M - 1, n_P, t) - P(n_G, n_M, n_P, t) \right]$$
$$+ k_m^- \left[(n_M + 1) P(n_G, n_M + 1, n_P, t) - n_M P(n_G, n_M, n_P, t) \right]$$
$$+ k_p n_M \left[P(n_G, n_M, n_P - 1, t) - P(n_G, n_M, n_P, t) \right]$$
$$+ k_p^- \left[(n_P + 1) P(n_G, n_M, n_P + 1, t) - n_P P(n_G, n_M, n_P, t) \right]$$
$$+ k_u \left(n_G^{tot} - n_G + 1 \right) P(n_G - 1, n_M, n_P - 1, t)$$
$$- k_u \left(n_G^{tot} - n_G \right) P(n_G, n_M, n_P, t)$$
$$+ k_b \left(n_G + 1 \right) \left(n_P + 1 \right) P(n_G + 1, n_M, n_P + 1, t)$$
$$- k_b n_G n_P P(n_G, n_M, n_P, t)$$

While the stochastic simulation algorithm and extensions provide a way to generate sample paths of copy numbers for a biochemical system, one needs to repeat many simulation runs to get an idea of the probability distribution in terms of its moments (mean and (co)variance), which become increasingly time-consuming and even impractical for larger systems. Therefore, attempts have been made toward continuous approximations of the CME [55, 65, 104, 123], including the following.

5.6 Continuous Approximations of the CME

It is imperative to rewrite the CME in an alternative notation more suited to a Taylor expansion. Using a negative-shift operator $\mathbb{E}_j$ for each reaction channel defined by its effect

$$\mathbb{E}_j f(n) = f(n + S_{\cdot j})$$

on an arbitrary scalar function $f(n)$ of s-vector n, the CME (5.40) can be written in the alternative form

$$\frac{\partial}{\partial t} P(n, t) = \sum_{j=1}^{r} \left(\mathbb{E}_j^{-1} - 1 \right) a_j(n) P(n, t). \tag{5.42}$$

5.6.1 Kramers–Moyal Expansion

Suppose the propensity $a_j(n)$ is a smooth function, and one is interested in solutions $P^c(n, t)$ that can be represented by a smooth function (defined for all real values of n, unlike the original probability distribution $P(n, t)$, which is defined only on integer values of n). These smoothness assumptions allow us to approximate the problem by means of a description in which n is treated as a continuous variable and thus replace the differential–difference equation by one partial differential equation. The operator $\mathbb{E}_j^{-1}$, acting only on smooth functions, may be replaced with a Taylor expansion,

$$\mathbb{E}_j^{-1} = \sum_{m=0}^{\infty} \frac{1}{m!} \left(-\sum_i S_{ij} \frac{\partial}{\partial n_i} \right)^m$$

$$= 1 - \sum_i S_{ij} \frac{\partial}{\partial n_i} + \frac{1}{2} \sum_{i,k} S_{ij} S_{kj} \frac{\partial^2}{\partial n_i \partial n_k} + \cdots. \tag{5.43}$$

Inserting values into the master equation yields the *Kramers–Moyal expansion*,

$$\frac{\partial}{\partial t} P^c(n, t) = \sum_{j=1}^{r} \sum_{m=1}^{\infty} \frac{1}{m!} \left(-\sum_i S_{ij} \frac{\partial}{\partial n_i} \right)^m a_j(n) P^c(n, t).$$

Fokker–Planck equation: Ignoring all the terms involving derivatives beyond the second in the above expansion, we obtain

$$\frac{\partial}{\partial t} P^c(n, t) = \sum_{j=1}^{r} \left(-\sum_i S_{ij} \frac{\partial}{\partial n_i} + \frac{1}{2} \sum_{i,k} S_{ij} S_{kj} \frac{\partial^2}{\partial n_i \partial n_k} \right) a_j(n) P^c(n, t),$$

which can be recast as the *Fokker–Planck equation* (FPE)

$$\frac{\partial}{\partial t} P^c(n,t) = -\frac{\partial}{\partial n}\left[A(n)P^c(n,t)\right] + \frac{1}{2}\frac{\partial^2}{\partial n \partial n^T}\left[B(n)P^c(n,t)\right], \qquad (5.44)$$

after recognizing the drift rate $A(n)$ and the diffusion rate $B(n)$.

Solving the nonlinear FPE (5.44) for the probability density is as difficult as the CME. Therefore, from an analytical point of view, the nonlinear FPE do not provide any significant advantage. However, linearising the propensity function around the mean [61], or using the Ω-expansion [75] (presented in the following section), the nonlinear FPE (5.44) can be reduced to the so-called "linear noise approximation" whose solution is a Gaussian distribution with a mean that is equal to the solution of the deterministic ODE model and a covariance matrix that obeys a linear ODE. This is the main drawback of linear noise approximation because, for system containing at least one bimolecular reactions, the mean of a stochastic model is not equal to the solution of deterministic ODEs, as shown next.

Relation to CLE: The CLE and FPE (5.44) are equivalent. It was shown in [55] that the probability density function $P^c(n,t)$ of the continuous approximation $N^c(t)$ described by the CLE (5.35) satisfies the FPE (5.44).

5.6.2 System-Size Expansion

The cutting of higher moments in the Kramers–Moyal expansion to get the Fokker–Planck approximation requires that the fluctuations, as measured by the standard deviations σ, be small. However, changes in copy numbers by chemical reactions are whole numbers, and there is no objective criterion of smallness of fluctuations in such a setting. This is true for any approximation method that requires small fluctuations. Therefore, one needs a systematic approximation method in the form of an expansion in powers of a small parameter. Only in that case does one have an objective measure for the size of the several terms. The expansion parameter must appear in the master equation and must govern the size of the fluctuations. The system-size parameter Ω is a potential choice. Again, we assume the propensity $a_j(n)$ to be a smooth function and look for solutions $P^c(n,t)$ that can be represented by a smooth function. It is then reasonable to approximate the problem by means of a description in which n is treated as a continuous variable. We can anticipate the way in which the solution $P^c(n,t)$ will depend on the system size Ω. The initial condition is

$$P^c(n,t) = \delta(n - n^0).$$

The initial copy number $X(0) = n^0 = \Omega x(0)$ is of order Ω. The Dirac delta function $\delta(n - n^0)$ is defined to be zero everywhere except at $n = n^0$, where it integrates to unity. One expects that at later times the distribution $P^c(n, t)$ has a sharp peak located at a value of order Ω, while its width will be of order $\sqrt{\Omega}$. In order words, it is assumed that the continuous approximation $N^c(t)$ of the process $N(t)$ fluctuates around a macroscopic trajectory of order Ω with a fluctuation of order $\Omega^{1/2}$. To express this formally, we set

$$N^c(t) = \Omega\phi(t) + \Omega^{1/2}\Xi(t), \tag{5.45}$$

where $\phi(t)$ is equal to the macroscopic concentration $x = n/\Omega$ for an infinitely large system size Ω and $\Xi(t)$ models the fluctuation of $N^c(t)$ around $\phi(t)$. A realization n of $N^c(t)$ is related to a realization ξ of $\Xi(t)$ by the same relation above:

$$n = \Omega\phi(t) + \Omega^{1/2}\xi.$$

The probability distribution $P^c(n, t)$ of $N^c(t)$ transforms into a probability distribution $\Pi(\xi, t)$ of $\Xi(t)$ according to

$$P^c(n, t) = P^c\left(\Omega\phi(t) + \Omega^{1/2}\xi, t\right) = \Pi(\xi, t). \tag{5.46}$$

The time derivative in the master equation is taken with constant n, that is,

$$\frac{dn}{dt} = \Omega\frac{d\phi}{dt} + \Omega^{1/2}\frac{d\xi}{dt} = 0 \quad \Longrightarrow \quad \frac{d\xi}{dt} = -\Omega^{1/2}\frac{d\phi}{dt}.$$

This result can be used in the differentiation of the probability distributions with respect to time to give

$$\frac{\partial P^c}{\partial t} = \frac{\partial \Pi}{\partial t} + \sum_{i=1}^{s}\frac{d\xi_i}{dt}\frac{\partial \Pi}{\partial \xi_i} = \frac{\partial \Pi}{\partial t} - \Omega^{1/2}\sum_{i=1}^{s}\frac{d\phi_i}{dt}\frac{\partial \Pi}{\partial \xi_i}. \tag{5.47}$$

Before we can compare this equation with the CME (5.42), we need to express the propensity function $a_j(n)$ in terms of the fluctuation ξ and translate the operator $\mathbb{E}_j$ so that it can be applied to functions of ξ. The propensity is related to the deterministic reaction rate $v_j(x)$ through (5.12):

$$a_j(n) = \Omega\left[v_j\left(\phi + \Omega^{-1/2}\xi\right) + \mathcal{O}\left(\Omega^{-1}\right)\right].$$

The operator $\mathbb{E}_j^{-1}$ that changes n to $n - S_{\cdot j}$, effectively changing the fluctuation ξ to $\xi - \Omega^{-1/2}S_{\cdot j}$, translates to $\mathbb{E}_j^{-\Omega^{-1/2}}$, which can be applied to functions of ξ. Now we can write the CME (5.42) so that the right side is a function of ξ

only:

$$\frac{\partial P^c}{\partial t} = \Omega \sum_{j=1}^{r} \left(\mathbb{E}_j^{-\Omega^{-1/2}} - 1 \right) \left[v_j \left(\phi + \Omega^{-1/2} \xi \right) + \mathcal{O} \left(\Omega^{-1} \right) \right] \Pi(\xi, t), \quad (5.48)$$

where the replacement of $P^c(n, t)$ with $\Pi(\xi, t)$ on the right follows from (5.46). The next step is the Taylor expansion, around ϕ, of $v_j(x)$ and the operator $\mathbb{E}_j^{-\Omega^{-1/2}}$ in several dimensions:

$$v_j \left(\phi + \Omega^{-1/2} \xi \right) = v_j(\phi) + \Omega^{-1/2} \sum_i \frac{\partial v_j}{\partial \phi_i} \xi_i + \mathcal{O} \left(\Omega^{-1} \right),$$

$$\mathbb{E}_j^{-\Omega^{-1/2}} = 1 - \Omega^{-1/2} \sum_i S_{ij} \frac{\partial}{\partial \xi_i}$$

$$+ \frac{1}{2} \Omega^{-1} \sum_{i,k} S_{ij} S_{kj} \frac{\partial^2}{\partial \xi_i \partial \xi_k} + \mathcal{O} \left(\Omega^{-3/2} \right),$$

where the latter follows from (5.43) by replacing n with ξ and S with $\Omega^{-1/2} S$. Inserting the above two expansions into (5.48) and then comparing the result with (5.47) leads to

$$\frac{\partial \Pi}{\partial t} - \Omega^{1/2} \sum_{i=1}^{s} \frac{d\phi_i}{dt} \frac{\partial \Pi}{\partial \xi_i}$$

$$= \sum_{j=1}^{r} \left(-\Omega^{1/2} \sum_i S_{ij} \frac{\partial}{\partial \xi_i} + \frac{1}{2} \sum_{i,k} S_{ij} S_{kj} \frac{\partial^2}{\partial \xi_i \partial \xi_k} + \mathcal{O} \left(\Omega^{-1/2} \right) \right)$$

$$\times \left[v_j(\phi) + \Omega^{-1/2} \sum_i \frac{\partial v_j}{\partial \phi_i} \xi_i + \mathcal{O} \left(\Omega^{-1} \right) \right] \Pi(\xi, t).$$

The terms of order $\Omega^{1/2}$ are proportional to the factors $\partial \Pi / \partial \xi_i$. It is possible to make the terms of each type cancel each other by choosing ϕ such that

$$\frac{d\phi_i}{dt} = \sum_{j=1}^{r} S_{ij} v_j(\phi), \quad (5.49)$$

and we see that the macroscopic law emerges as the lowest approximation in
the Ω expansion. Comparing terms of order Ω^0 yields

$$\frac{\partial \Pi}{\partial t} = \sum_{j=1}^{r} \left(-\sum_{i,k} S_{ij} \frac{\partial v_j}{\partial \phi_k} \frac{\partial (\xi_k \Pi)}{\partial \xi_i} + \frac{1}{2} v_j(\phi) \sum_{i,k} S_{ij} S_{kj} \frac{\partial^2 \Pi}{\partial \xi_i \partial \xi_k} \right) .$$

Introducing matrices G and H with elements

$$G_{ik} = \sum_{j=1}^{r} S_{ij} \frac{\partial v_j}{\partial \phi_k} \quad \text{and} \quad H_{ik} = \sum_{j=1}^{r} S_{ij} S_{kj} v_j(\phi), \tag{5.50}$$

the above differential equation can be written as

$$\frac{\partial \Pi}{\partial t} = -\sum_{i,k} G_{ik} \frac{\partial (\xi_k \Pi)}{\partial \xi_i} + \frac{1}{2} \sum_{i,k} H_{ik} \frac{\partial^2 \Pi}{\partial \xi_i \partial \xi_k}, \tag{5.51}$$

which is a *linear Fokker–Planck equation* with coefficient matrices G and H
that depend on time through macroscopic concentration $\phi(t)$. The temporal
dynamics of the mean fluctuations $\langle \Xi_i \rangle$ can be obtained by multiplying (5.51)
by ξ_i and integrating over all values ξ:

$$\frac{d \langle \Xi_i \rangle}{dt} = \sum_{k} G_{ik} \langle \Xi_i \rangle . \tag{5.52}$$

By a similar procedure, the temporal dynamics of the second moments $\langle \Xi_i \Xi_k \rangle$
of pairs of fluctuations can be obtained by multiplying (5.51) by $\xi_i \xi_k$ and
integrating over all values ξ:

$$\frac{d \langle \Xi_i \Xi_k \rangle}{dt} = \sum_{l} G_{il} \langle \Xi_l \Xi_k \rangle + \sum_{l} G_{kl} \langle \Xi_i \Xi_l \rangle + H_{ik} . \tag{5.53}$$

It is convenient to consider instead of these moments the covariances

$$\langle \delta \Xi_i \delta \Xi_k \rangle = \langle \Xi_i \Xi_k \rangle - \langle \Xi_i \rangle \langle \Xi_k \rangle .$$

With the aid of (5.52) one finds that they satisfy the same equation (5.53),
but vanish at $t = 0$. In matrix notation,

$$\frac{d \langle \delta \Xi \delta \Xi^T \rangle}{dt} = G \langle \delta \Xi \delta \Xi^T \rangle + \langle \delta \Xi \delta \Xi^T \rangle G^T + H . \tag{5.54}$$

With the mean and covariance determined by (5.52)–(5.54), the solution of
the linear FPE (5.51) can be shown to be a multivariate Gaussian [77]:

$$\Pi(\xi, t) = (2\pi)^{-s/2} \left| \langle \delta\Xi\delta\Xi^T \rangle \right|^{-1/2} \exp \left(-\frac{1}{2} \xi^T \langle \delta\Xi\delta\Xi^T \rangle^{-1} \xi \right). \qquad (5.55)$$

The proposed transformation (5.45) together with (5.49) and (5.51) forms the
so-called *linear noise approximation* (LNA) [75].

Since the LNA does not include terms of order higher than Ω^0, the
same could have been obtained by applying the method of Ω-expansion to
the nonlinear FPE (5.44).

Tracking the moments: The analytical approximations discussed in this
chapter do not allow direct tracking of the mean and (co)variance, which,
in general, are coupled. This coupling is not obvious in CLE and FPE, and
ignored by LNA and conventional ODE models. The next chapter presents
the 2MA approach, which has a direct representation of the first two moments
and the coupling between them.

Problems

5.1. Revisit the metabolite network in Exercise 2.6,

$$2X_1 \xrightarrow{k_1} X_2, \quad X_2 + X_3 \xrightarrow{k_2} X_4.$$

In that continuous description you directly worked with the metabolite concentrations. Take the same rate constants and initial abundances, but this
time you have to deal with copy numbers!

1. The reaction rate constants k_1 and k_2 have units involving nanomolars
   and cannot be used in discrete stochastic simulations. Compute the
   conversion rate constants $\hat{k}_1$ and $\hat{k}_2$. Use $5\,\mu m^3$ for the volume V.

2. Give the expressions for the reaction propensities a_1 and a_2 in terms of
   the copy number $n = (n_1, n_2, n_3, n_4)^T$.

3. Implement these expressions in Matlab to compute propensities at any
   state n.

4. Instead of computing propensities directly, since both the reactions
   are elementary, you could call the function makePropensity with
   arguments c (the stochastic rate constant) and S (the reactant stoichiometry). For that, the conversion rate constants must be translated

to stochastic rate constants c_1 and c_2. Rewrite the above code by calling `makePropensity`.

5.2. For the metabolite network in Exercise 2.6, you have obtained Matlab representations of the stoichiometry matrix S and the reaction propensity function $a(\cdot)$. In this exercise, you will use these two in discrete stochastic simulations based on Gillespie's SSA.

1. Generate a sample path of the metabolite copy numbers for 500 seconds. You could make use of the first handle returned by the function `makeSSA`, which requires as its arguments the stoichiometry matrix S and a function handle for the propensity, which you already have. Repeat the simulation for a different volume, say $2\,\mu m^3$. What differences do you observe among repeated simulation runs for the two volume? Explain your observations.

2. To get an idea of the average behavior and fluctuations around it, one could plot the ensemble mean together with mean±SD. Complete the following code by adding code to compute the mean and variance from 100 simulation runs. You could use the second handle returned by the function `makeSSA` for that purpose.

```
% mu = ?; % mean
% sd = ?; % standard deviation
figure
ti = linspace(0,tt(end),20).';
for i=1:4
    nni = interp1q(tt, [mu(:,i) sd(:,i)], ti);
    errorbar(ti,nni(:,1),nni(:,2));
    hold on;
end
hold off
```

3. The plots in the last part do not indicate how the fluctuations scale with the average copy numbers. Plot the standard deviation divided by the square root of the mean and check it for different volumes. Check whether the plots confirm the inverse-square-root relationship with the copy numbers.

4. Plot the endpoint (or terminal) histogram for all species for different volumes. How does the histogram change with the system size?

5.3. For the same metabolite network in Exercise 2.6, you are going to use the Matlab representations of the stoichiometry matrix S and the reaction propensity function $a(\cdot)$ in continuous stochastic simulations based on CLE.

Call the Matlab function `simCLE` (in the main text) to generate a sample path of the metabolite copy numbers for 500 seconds. Compare your observations with those from discrete stochastic simulations. Repeat the other tasks in the previous exercise for continuous stochastic simulations.

5.4. Recall the repressilator in Exercise 2.7, for which you worked out the stoichiometry matrix S and the reaction rate vector $v(x)$. Now you have to revisit it from a stochastic perspective. To keep things simple, assume that one unit of concentration corresponds to one molecule in the network.

1. Write down the expressions for channelwise propensities $a_j(n)$ in terms of species copy number n.

2. Complete the following script to compute, using SSA, and plot the protein levels for 50 time units:

```
% parameters
a0 = 0.25; a1 = 250; b = 5; h = 2.1;
% stoichiometry matrix
Splus = diag(ones(6,1));
S = [-Splus Splus];
% prop = @(n) ?; % propensity
[ssa,ensem] = makeSSA(S,prop);
tmax = 50; dt = 0.1 % time scale and steps
n0 = [0 0 0 4 0 15]'; % initial copy number
[t,n] = ssa(n0,tmax,dt); % SSA run
plot(t,n(:,4:6)) % plot protein levels
```

Do you see oscillations in the protein levels? Play with the parameter values and initial conditions to see whether you always get oscillations.

5.5. Repeat the second task in the previous exercise using continuous stochastic simulations based on the CLE. Call `simCLE` for that purpose. Compare your results with the discrete simulations. Do you still see oscillations in the protein levels?

Chapter 6

The 2MA Approach

This chapter develops a compact form of the 2MA equations—a system of ODEs for the dynamics of the mean and (co)variance of the continuous-time discrete-state Markov process that models a biochemical reaction system by the CME. This is an extension of previous derivations, taking into account relative concentrations and nonelementary reactions. The compact form, obtained by careful selection of notation, allows for an easier interpretation.

The 2MA approach allows a representation of the coupling between the mean and (co)variance. The traditional Langevin approach is based on the assumption that the time rate of change of abundance (copy number or concentration) or the flux of a component can be decomposed into a deterministic flux and a Langevin noise term, which is a Gaussian (white noise) process with zero mean and amplitude determined by the system dynamics. This separation of noise from the system dynamics may be a reasonable assumption for *external noise* that arises from the interaction of the system with other systems (such as the environment), but cannot be assumed for internal noise that arises from within the system [13, 30, 74, 117, 128, 141]. As categorically discussed in [76], internal noise is not something that can be isolated from the system, because it results from the discrete nature of the underlying molecular events. Any noise term in the model must be derived from the system dynamics and cannot be presupposed in an ad hoc manner. However, the CLE does not suffer from the above criticism, because Gillespie [56] derived it from the CME description. The CLE allows much faster simulations compared to the exact stochastic simulation algorithm (SSA) [52] and its variants. The CLE is a stochastic differential equation (dealing directly with random variables rather than moments) and has no direct way of representing the mean and (co)variance and the coupling between the two. That does not imply that CLE ignores the coupling as does the LNA, which has the same mean as the solution of the deterministic model.

6.1 Absolute and Relative Concentrations

The concentration $X(t)$ is usually defined as the copy number $N(t)$ divided by the system size parameter Ω. In other words,

$$N(t) = \Omega X(t).$$

However, for some systems it is more appropriate to introduce a different scaling parameter Ω_i for each component i if the copy numbers N_i differ in magnitude to keep X_i of the same order $\mathcal{O}(1)$. In such cases, the concentration will be defined as a relative concentration

$$X_i = \frac{N_i}{C_i \Omega},$$

that is, the absolute concentration N_i/Ω divided by a characteristic concentration C_i. In that case, each scaling parameter can be expressed as $\Omega_i = C_i \Omega$. Unless otherwise stated, concentration X_i will be interpreted in the absolute sense.

Often we are interested in the first two moments of the probability distribution. The first moment is the mean vector $\langle N(t) \rangle$ of copy numbers, defined elementwise by

$$\langle N_i(t) \rangle = \sum_n n_i P(n, t),$$

the ith mean copy number. The second central moment is the covariance matrix $\langle \delta N \delta N^T \rangle$ defined elementwise by

$$\langle \delta N_i \delta N_k \rangle = \left\langle \left(N_i - \langle N_i \rangle \right) \left(N_k - \langle N_k \rangle \right) \right\rangle,$$

the covariance between N_i and N_k. When obvious from the context, we will leave out dependence on time, as in the above.

We are also interested in the mean concentration vector

$$\langle X \rangle = \frac{\langle N \rangle}{\Omega}$$

and the concentration covariance matrix

$$\langle \delta X \delta X^T \rangle = \frac{\langle \delta N \delta N^T \rangle}{\Omega^2}.$$

The diagonal elements of the covariance matrix are the variances

$$\langle \delta N_i^2 \rangle = \langle \delta N_i \delta N_i \rangle, \quad \langle \delta X_i^2 \rangle = \langle \delta X_i \delta X_i \rangle .$$

6.2 Dynamics of the Mean

Taking expectation on both sides of (2.3) gives the mean copy number,

$$\langle N(t) \rangle = N(0) + S \langle Z(t) \rangle .$$

Taking the time derivative and employing (5.9) yields

$$\frac{d \langle N \rangle}{dt} = S \langle a(N) \rangle = \langle A(N) \rangle, \tag{6.1}$$

where we have recognized the drift rate $A(n) = S a(n)$ of the copy-number process $N(t)$ defined in (5.29). Dividing by Ω gives the system of ODEs for the mean concentration:

$$\frac{d \langle X \rangle}{dt} = \langle f(X) \rangle, \tag{6.2}$$

where $f(x) = S v(x)$ is the drift rate of the concentration process $X(t)$. In general (to account for relative concentrations), the drift rate can be defined componentwise by

$$f_i(x) = \frac{1}{C_i} \sum_{j=1}^{r} S_{ij} v_j(C \odot x). \tag{6.3}$$

Here C is the s-vector of characteristic concentrations C_i, and the binary operation $\odot$ denotes the elementwise product of two arrays. The two drift rates are related by $A(n) = \Omega f(x)$ for absolute concentrations and, componentwise, by $A_i(n) = \Omega C_i f_i(x)$ for relative concentrations. The above two systems of ODEs for the mean abundance should be compared to the analogous chemical kinetic equations (2.16) for the deterministic abundance.

It is interesting to note that (6.1) is a direct consequence of mass conservation (2.2) and definition of propensity because we have not referred to the CME (which is the usual procedure) during our derivation.

Example 6.1 (Standard modification) Following from the state transitions in Figure 5.19 for the standard modification (2.5), the reaction propensities are linear,

$$v_1 = k_w x, \quad v_2 = (1 - x) k_u,$$

giving the linear drift rate

$$f(x) = -v_1 + v_2 = k_u - (k_w + k_u)\, x\,.$$

The mean copy number thus satisfies

$$\frac{\mathrm{d}\,\langle X \rangle}{\mathrm{d}t} = \langle f(X) \rangle = k_u - (k_w + k_u)\,\langle X \rangle\,,$$

which is the same as the deterministic ODE (2.21). In general, the mean of a system composed solely of reactions of order zero and/or one is the same as the solution of the corresponding deterministic ODE because of linear propensities.

Example 6.2 (Lotka–Volterra model) Following the state transitions in Figure 5.24 for the reaction scheme (2.7) of the Lotka–Volterra model, the reaction propensities are given by

$$a_1 = \hat{k}_1 n_{\mathrm{A}} n_1, \quad a_2 = \hat{k}_2 n_1 n_2, \quad a_3 = \hat{k}_3 n_2,$$

giving the drift rate

$$A(n) = \begin{bmatrix} a_1 - a_2 \\ a_2 - a_3 \end{bmatrix} = \begin{bmatrix} \hat{k}_1 n_{\mathrm{A}} n_1 - \hat{k}_2 n_1 n_2 \\ \hat{k}_2 n_1 n_2 - \hat{k}_3 n_2 \end{bmatrix}.$$

The time derivative of the mean copy number is then the mean drift rate $\langle A(N) \rangle$:

$$\frac{\mathrm{d}\,\langle N_1 \rangle}{\mathrm{d}t} = \hat{k}_1 n_{\mathrm{A}}\,\langle N_1 \rangle - \hat{k}_2\,\langle N_1 N_2 \rangle\,,$$

$$\frac{\mathrm{d}\,\langle N_2 \rangle}{\mathrm{d}t} = \hat{k}_2\,\langle N_1 N_2 \rangle - \hat{k}_3\,\langle N_2 \rangle\,.$$

Since $\langle N_1 N_2 \rangle \neq \langle N_1 \rangle \langle N_2 \rangle$, the mean $\langle N \rangle$ is not the same as the solution of the deterministic ODE (2.23). In general, the mean of a system containing second- and/or higher-order reactions is not the same as the solution of the corresponding deterministic ODE because of nonlinear propensities. Hence the deterministic model cannot be used, in general, to describe the correct mean.

In general, the mean drift rate $\langle A(N) \rangle$ involves the unknown probability distribution $P(n,t)$. In other words, it not only depends on the mean itself, but also involves higher-order moments, and therefore (6.2) is, in general, not closed in $\langle N \rangle$. Suppose the reaction propensities $a_j(n)$ are smooth functions and that central moments $\langle (N - \langle N \rangle)^m \rangle$ of order higher than $m = 2$ can be ignored. Then the elements of drift rate $A(n) = S\,a(n)$ are also smooth functions and allow a Taylor expansion. The Taylor expansion of $A_i(n)$ around the mean $\langle N \rangle$ is

$$A_i(n) = A_i\left(\langle N \rangle\right) + \frac{\partial A_i}{\partial n^T}\left(n - \langle N \rangle\right) + \frac{1}{2}\left(n - \langle N \rangle\right)^T \frac{\partial^2 A_i}{\partial n \partial n^T}\left(n - \langle N \rangle\right) + \cdots .$$

All the partial derivatives with respect to the state n are evaluated at $n = \langle N \rangle$. The first-order partial derivative here is the ith row of the Jacobian $\frac{\partial A}{\partial n^T}$ of the drift rate A. The second-order partial derivative is the Hessian $\frac{\partial^2 A_i}{\partial n \partial n^T}$ of the componentwise drift rate A_i. The expectation of the second term on the right is zero. Ignoring terms (moments) of order higher than two, the ODE (6.1) can be approximated componentwise by

$$\frac{d\langle N_i \rangle}{dt} = A_i\left(\langle N \rangle\right) + \frac{1}{2}\frac{\partial^2 A_i}{\partial n \partial n^T} : \langle \delta N \delta N^T \rangle . \tag{6.4}$$

Here the binary operation : denotes the Frobenius inner product, the sum of products of the corresponding elements, between two matrices. The last term on the right can be interpreted as the contribution from fluctuations in the drift rate. Note that this term has been derived from the CME instead of being assumed like external noise. This shows that knowledge of fluctuations (even if small) is important for a correct description of the mean. This also indicates an advantage of the stochastic framework over its deterministic counterpart: starting from the same assumptions and approximations, the stochastic framework allows us to see the influence of fluctuation on the mean. Note that the above equation is exact for systems in which no reaction has an order higher than two, because then third and higher derivatives of propensity are zero.

Dividing the above ODE by the system size Ω, we can arrive at the corresponding ODE for componentwise concentration,

$$\frac{d\langle X_i \rangle}{dt} = f_i\left(\langle X \rangle\right) + \frac{1}{2}\frac{\partial^2 f_i}{\partial x \partial x^T} : \langle \delta X \delta X^T \rangle , \tag{6.5}$$

where all the partial derivatives with respect to the state x are evaluated at $x = \langle X \rangle$ and f_i is the componentwise drift rate. The above result follows from the following obvious relations in state $n = \Omega x$:

$$\langle N \rangle = \Omega \langle X \rangle , \quad \langle \delta N \delta N^T \rangle = \Omega^2 \langle \delta X \delta X^T \rangle , \quad A(n) = \Omega f(x) .$$

The form of the above ODE will not change in dealing with relative concentrations because then we would employ the relations

$$\langle N_i \rangle = \Omega C_i \langle X_i \rangle, \quad \langle \delta N_i \delta N_k \rangle = \Omega^2 C_i C_k \langle \delta X_i \delta X_k \rangle, \quad A_i(n) = \Omega C_i f_i(x),$$

and thus the characteristic concentrations would cancel out during the simplification process from (6.4) to (6.5). Remember that all the derivatives with respective state vectors n and x are evaluated at $n = \langle N \rangle$ and $x = \langle X \rangle$, respectively.

6.3 Dynamics of the (Co)variance

Before we can see how the (co)variances $\langle N_i N_k \rangle$ evolve in time, let us multiply the CME by $n_i n_k$ and sum over all n,

$$\sum_n n_i n_k \frac{\mathrm{d}P(n,t)}{\mathrm{d}t} = \sum_n n_i n_k \sum_j \left[a_j(n - S_{\cdot j})P(n - S_{\cdot j}, t) - a_j(n)P(n,t) \right]$$

$$= \sum_{n,j} \left[(n_i + S_{ij})(n_k + S_{kj}) - n_i n_k \right] a_j(n)P(n,t)$$

$$= \sum_n \sum_j \left(n_k S_{ij} + n_i S_{kj} + S_{ij} S_{kj} \right) a_j(n)P(n,t),$$

where dependence on time is implicit for all variables except n and s. Recognizing sums of probabilities as expectations yields

$$\frac{\mathrm{d} \langle N_i N_k \rangle}{\mathrm{d}t} = \langle N_k A_i(N) \rangle + \langle N_i A_k(N) \rangle + \langle B_{ik}(N) \rangle,$$

where

$$B_{ik}(n) = \sum_{j=1}^{r} S_{ij} S_{kj} a_j(n)$$

forms the (i,k)th element of the $s \times s$ *diffusion coefficient* $B(n)$ defined earlier, in (5.31). The relation

$$\langle \delta N \delta N^T \rangle = \langle N N^T \rangle - \langle N \rangle \langle N \rangle^T$$

can be utilized to yield

$$\frac{\mathrm{d} \langle \delta N_i \delta N_k \rangle}{\mathrm{d}t} = \langle (N_k - \langle N_k \rangle) A_i(N) \rangle + \langle (N_i - \langle N_i \rangle) A_k(N) \rangle + \langle B_{ik}(N) \rangle \quad (6.6)$$

for the copy-number covariance. Dividing through by Ω^2 gives the analogous system of ODEs for concentration covariance:

$$\frac{d \langle \delta X_i \delta X_k \rangle}{dt} = \langle (X_k - \langle X_k \rangle) f_i(X) \rangle + \langle (X_i - \langle X_i \rangle) f_k(X) \rangle + \frac{\langle g_{ik}(X) \rangle}{\Omega}, \quad (6.7)$$

where $g_{ik}(x) = \sum_{j=1}^r S_{ij} S_{kj} v_j(x)$ for the case of absolute concentrations, and in general (to account for relative concentrations),

$$g_{ik}(x) = \frac{1}{C_i C_k} \sum_{j=1}^r S_{ij} S_{kj} v_j(C \odot x), \quad (6.8)$$

an element of the normalized diffusion coefficient g in terms of concentrations. Let us start with the component form of (6.6). The argument of the first expectation on the right has Taylor expansion

$$A_i(n)\left(n_k - \langle N_k \rangle\right) = A_i\left(\langle N \rangle\right)\left(n_k - \langle N_k \rangle\right) + \frac{\partial A_i}{\partial n^T}\left(n - \langle N \rangle\right)\left(n_k - \langle N_k \rangle\right) + \cdots .$$

The expectation of the first term on the right is zero. Ignoring moments of order higher than two, the first expectation in (6.6) is then

$$\left\langle (N_k - \langle N_k \rangle) A_i(N) \right\rangle = \frac{\partial A_i}{\partial n^T} \langle \delta N \delta N_k \rangle .$$

By a similar procedure, the second expectation in (6.6) is

$$\left\langle (N_i - \langle N_i \rangle) A_k(N) \right\rangle = \langle \delta N_i \delta N^T \rangle \frac{\partial A_k}{\partial n},$$

correct to second-order moments. The element $B_{ik}(n)$ of the diffusion coefficient has Taylor expansion

$$B_{ik}(n) = B_{ik}\left(\langle N \rangle\right) + \frac{\partial B_{ik}}{\partial n^T}\left(n - \langle N \rangle\right) + \frac{1}{2}\left(n - \langle N \rangle\right)^T \frac{\partial^2 B_{ik}}{\partial n \partial n^T}\left(n - \langle N \rangle\right) + \cdots .$$

Taking termwise expectation, and ignoring third and higher-order moments yields

$$\langle B_{ik}(N) \rangle = B_{ik}\left(\langle N \rangle\right) + \frac{1}{2}\frac{\partial^2 B_{ik}}{\partial n \partial n^T} : \langle \delta N \delta N^T \rangle .$$

Summing up the three expectations above gives the ODE

$$\frac{\mathrm{d}\langle \delta N_i \delta N_k\rangle}{\mathrm{d}t} = \frac{\partial A_i}{\partial n^T}\langle \delta N \delta N_k\rangle + \langle \delta N_i \delta N^T\rangle \frac{\partial A_k}{\partial n}$$
$$+ B_{ik}\big(\langle N\rangle\big) + \frac{1}{2}\frac{\partial^2 B_{ik}}{\partial n \partial n^T} : \langle \delta N \delta N^T\rangle \qquad (6.9)$$

for the componentwise covariances. The Jacobian $\partial f/\partial n^T$ reflects the dynamics for relaxation (dissipation) to the steady state and the (Taylor approximation to second order of) diffusion coefficient B reflects the randomness (fluctuations) of the individual events [120]. These terms are borrowed from the *fluctuation–dissipation theorem* (FDT) [78, 89], which has the same form as (6.9). Recall that (6.9) is exact for systems that contain only zero- and first-order reactions because in that case the propensity is already linear.

Dividing the above ODE by Ω^2 will lead to the corresponding ODE for the pairwise concentration covariance:

$$\frac{\mathrm{d}\langle \delta X_i \delta X_k\rangle}{\mathrm{d}t} = \frac{\partial f_i}{\partial x^T}\langle \delta X \delta X_k\rangle + \langle \delta X_i \delta X^T\rangle \frac{\partial f_k}{\partial x}$$
$$+ \frac{1}{\Omega}\left[g_{ik}\big(\langle X\rangle\big) + \frac{1}{2}\frac{\partial^2 g_{ik}}{\partial x \partial x^T} : \langle \delta X \delta X^T\rangle\right]. \qquad (6.10)$$

Here we see that steady-state concentration (co)variance is inversely proportional to the system size Ω, as one would expect. In other words, the steady-state noise, as measured by the square root of (co)variance, is inversely proportional to the square root of the system size, as one would expect. The above result follows from the following obvious relations in state $n = \Omega x$:

$$\langle N\rangle = \Omega \langle X\rangle, \qquad\qquad \langle \delta N \delta N^T\rangle = \Omega^2 \langle \delta X \delta X^T\rangle,$$
$$A(n) = \Omega f(x), \qquad\qquad B(n) = \Omega^2 g(x).$$

As we noted in the last section, the form of the above ODE will not change in dealing with relative concentrations, because then we would employ the following relations:

$$\langle N_i\rangle = \Omega C_i \langle X_i\rangle, \qquad\qquad \langle \delta N_i \delta N_k\rangle = \Omega^2 C_i C_k \langle \delta X_i \delta X_k\rangle,$$
$$A_i(n) = \Omega C_i f_i(x), \qquad\qquad B_{ik}(n) = \Omega^2 C_i C_k g_{ik}(x)$$

and thus the characteristic concentrations would cancel out for the third term during the simplification process from (6.9) to (6.10).

6.3.1 Potential Confusions

It is instructive to point out similarities and differences of the 2MA approach to other approaches.

The Gaussian approximation: The fact that the 2MA approach tracks only the first two central moments may misguide the reader to equate the 2MA approach to a Gaussian approximation is characterized only by its first two central moments. However, the 2MA approach ignored all the higher-order central moments, whereas for the Gaussian distribution only the odd higher-order central moments are zero. The even higher-order (central) moments of a Gaussian distribution can be expressed as the standard deviation raised to the corresponding order.

The CLE: Consider the system in state $n = \Omega x$. The drift rate $A(n)$ (defined by (5.29) and appearing in the 2MA equations) appears as the coefficient of dt of the first term on the right of the CLE (5.35) and (5.39). Moreover, the diffusion coefficient B (defined by (5.31) and appearing in the 2MA equations) has its matrix square root $B^{1/2}$ as a coefficient of the Wiener increment in the CLE (5.39). However, while the CLE essentially approximates the Markov process $N(t)$ by a Gaussian random process, the 2MA simply ignores the third and higher-order central moments.

Alternative names: In [58], the 2MA framework is developed under the name "mass fluctuation kinetics" for biochemical networks composed of elementary reactions. Another instance of the 2MA is proposed in [61, 62] under the names "mean-field approximation" and "statistical chemical kinetics."

6.4 Outline of the 2MA

We can summarize the 2MA method, in terms of copy numbers, by the following steps:

1. Assign propensity $a_j(n)$ to each reaction channel.

2. Construct elements $A_i(n)$ of the drift rate according to (5.29) and the partial derivatives $\frac{\partial A_i}{\partial n^T}$ and $\frac{\partial^2 A_i}{\partial n \partial n^T}$ for each species.

3. Construct elements $B_{ik}(n)$ of the diffusion coefficient according to (5.31) and the partial derivative $\frac{\partial^2 B_{ik}}{\partial n \partial n^T}$ for each pair of species.

4. Construct sums $\frac{\partial^2 A_i}{\partial n \partial n^T} : \langle \delta N \delta N^T \rangle$ of elementwise products between the Hessian of A_i and covariance matrix, for each species.

5. Construct the scalar (dot) products $\frac{\partial A_i}{\partial n}^T \langle \delta N \delta N_k \rangle$ and $\langle \delta N_i \delta N^T \rangle \frac{\partial A_k}{\partial n}$ for each pair of species.

6. Construct sums of elementwise products, $\frac{\partial^2 B_{ik}}{\partial n \partial n}^T : \langle \delta N \delta N^T \rangle$, between the Hessian of B_{ik} and covariance matrix for each pair of species.

7. Insert the expressions obtained so far in (6.4) and (6.9) to obtain the 2MA equations.

Remember that any derivatives with respective to state n must be evaluated at $n = \langle N \rangle$. The procedure for the 2MA equations in terms of concentrations is the same except for appropriate replacements such as N with X and so on.

Matlab implementation: The steps outlined above for the 2MA method can be implemented in Matlab. The implementation in M-code 6.1 is based on the symbolic toolbox. Here the output hOut returned by the main function make2MA is a function handle that can be used in any Matlab solver. Note that this function handle can be used in an ODE solver as if it were a function of the form $f(t, y)$ with two arguments: time t and a vector y that corresponds to the state n in the following way. The first s elements form the mean vector $\langle N \rangle$, and the remaining $s(s+1)/2$ elements form the lower triangular submatrix, read columnwise, of the covariance matrix $\langle \delta N \delta N^T \rangle$.

Example 6.3 (Standard modification) Let us practice the procedure outlined above to construct the 2MA equations for the standard modification scheme (2.5) with the associated state transitions in Figure 5.19.

1. Assign propensity $a_j(n)$ to each reaction channel:

$$a_1 = k_w n, \quad a_2 = \left(n^{\text{tot}} - n\right) k_u .$$

2. Construct the drift rate,

$$A(n) = a_1 - a_2 = -k_w n + \left(n^{\text{tot}} - n\right) k_u,$$

(according to(5.29)) and its partial derivatives:

$$\frac{\partial A}{\partial n} = -\left(k_w + k_u\right), \quad \frac{\partial^2 A}{\partial n^2} = 0 .$$

3. Construct the diffusion coefficient $B_{ik}(n)$ according to (5.31) and the partial derivative $\frac{\partial^2 B_{ik}}{\partial n \partial n}^T$ for each pair of species:

$$B(n) = a_1 + a_2 = k_w n + \left(n^{\text{tot}} - n\right) k_u, \quad \frac{\partial^2 B}{\partial n^2} = 0 .$$

M-code 6.1 make2MA: implements the 2MA method. Requires the symbolic toolbox.

```
function hOut = make2ma(S,a)
s = size(S,1); % number of species
n = sym('n%d', [s 1]); % copy number
C = sym('C%d%d', [s s]); % covariance matrix
aofn = a(n); % propensity
A = S*aofn; % drift rate
B = S*diag(aofn)*S.'; % diffusion coefficient
if s==1
    odearg = [n C].';
    d1Adn = diff(A, n); % Jacobian
    d2Adn2 = diff(d1Adn,n); % Hessian
    dndt = A +  0.5*d2Adn2.*C;
    dCdt = 2*d1Adn*C + B + 0.5*d2Adn2.*C;
else
    C = tril(C) + tril(C,-1).';
    onesxs = ones(s);
    ind = find(tril(onesxs));
    odearg = [n; C(ind)];
    d1Adntr = jacobian(A, n); % Jacobian
    d1Atrdn = d1Adntr.';
    dndt = A;
    for i=1:s
        d2Aidn2 = jacobian(d1Atrdn(:,i),n); % Hessian
        dndt(i) = dndt(i) + 0.5*sum(d2Aidn2(:).*C(:));
    end
    dCdt = d1Adntr*C;
    dCdt = dCdt + dCdt.';
    for idx=ind(:).'
        d1Bijdn = jacobian(B(idx), n).';
        d2Bijdn2 = jacobian(d1Bijdn,n);
        dCdt(idx) = dCdt(idx) + ...
            B(idx) + 0.5*sum(d2Bijdn2(:).*C(:));
    end
    dCdt = dCdt(ind);
end
hOut = matlabFunction( ...
    [dndt; dCdt], 'vars', {sym('t'),odearg} );
```

4. Construct sums $\frac{\partial^2 A_i}{\partial n \partial n^T} : \langle \delta N \delta N^T \rangle$ of elementwise products between the Hessian of A_i and covariance matrix for each species:

$$\frac{\partial^2 A}{\partial n^2} \langle \delta N^2 \rangle = 0 .$$

5. Construct the scalar (dot) products $\frac{\partial A_i}{\partial n^T} \langle \delta N \delta N_k \rangle$ and $\langle \delta N_i \delta N^T \rangle \frac{\partial A_k}{\partial n}$ for each pair of species:

$$\frac{\partial A}{\partial n} \langle \delta N^2 \rangle = - (k_w + k_u) \langle \delta N^2 \rangle .$$

6. Construct sums $\frac{\partial^2 B_{ik}}{\partial n \partial n^T} : \langle \delta N \delta N^T \rangle$ of elementwise products between the Hessian of B_{ik} and covariance matrix for each pair of species:

$$\frac{\partial^2 B}{\partial n^2} \langle \delta N^2 \rangle = 0 .$$

7. Insert the expressions obtained so far in (6.4) and (6.9) to obtain the 2MA equations

$$\frac{\mathrm{d} \langle N \rangle}{\mathrm{d}t} = - (k_w + k_u) \langle N \rangle + k_u n^{\mathrm{tot}},$$

$$\frac{\mathrm{d} \langle \delta N^2 \rangle}{\mathrm{d}t} = -2 (k_w + k_u) \langle \delta N^2 \rangle + (k_w - k_u) \langle N \rangle + k_u n^{\mathrm{tot}} . \tag{6.11}$$

We see that the growth of variance is influenced by the mean through the rate term. With a rise in the mean, the growth of variance speeds up if $k_w > k_u$, slows down if $k_w < k_u$, and is not influenced if $k_w = k_u$. This is illustrated in Figure 6.1, which plots the standard deviation (SD) and the coefficient of variation (CV) for four pairs of parameter values with the same sum $k_w + k_u = 4$. It is interesting to note that the transient overshoot of the SD is not shared by the CV. To get a qualitative idea about possible stochastic realizations, the mean and the band of one standard deviation around it are plotted in Figure 6.2 for the same three pairs of parameter values.

In nondimensional time $\tau = (k_w + k_u)t$, the above pair of ODEs takes the form

$$\frac{\mathrm{d} \langle N \rangle}{\mathrm{d}\tau} = - \langle N \rangle + \left(\frac{k_u}{k_w + k_u} \right) n^{\mathrm{tot}},$$

$$\frac{\mathrm{d} \langle \delta N^2 \rangle}{\mathrm{d}\tau} = -2 \langle \delta N^2 \rangle + \left(\frac{k_w - k_u}{k_w + k_u} \right) \langle N \rangle + \left(\frac{k_u}{k_w + k_u} \right) n^{\mathrm{tot}} .$$

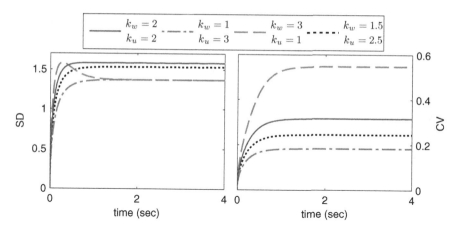

Figure 6.1 Time courses, for the standard modification, of standard deviation (SD), on the left, and the coefficient of variation (CV), on the right. The parameter pairs (k_w, k_u) in $\sec^{-1}$ have been selected to satisfy $k_w + k_u = 4$. The total number of protein molecules was chosen to be $n^{\text{tot}} = 10$, initially all unmodified, that is, $N(0) = 10$.

Now we can see that experimental data on both the mean and variance are needed for identifiability of both the parameters. The time-course measurements of mean alone provide information about one fraction $k_u/(k_w+k_u)$ only. To get information about the other fraction $(k_w-k_u)/(k_w+k_u)$, we need time-course measurements of variance as well.

Here we have used the SD and CV as measures of noise. In addition to these two, other alternative measures of noise, including the Fano factor $F = \langle \delta N^2 \rangle / \langle N \rangle$ and the noise-to-signal ratio $\zeta = \langle \delta N^2 \rangle / \langle N \rangle^2$, are discussed in [117].

Example 6.4 (Heterodimerization) For the heterodimerization reaction scheme (2.6) with state transitions in Figure 5.22, the 2MA equations are obtained by the same procedure outlined earlier.

1. Assign kinetic rate $v_j(x)$ to each reaction channel:

$$v_1 = k_1 (q_1 - x)(q_2 - x), \quad a_2 = k_2 x.$$

2. Construct the drift rate,

$$f = v_1 - v_2 = k_1 (q_1 - x)(q_2 - x) - k_2 x,$$

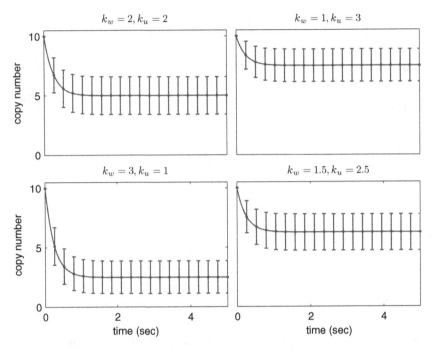

Figure 6.2 Time courses of mean and mean±SD for the standard modification for four parameter pairs (k_w, k_u) in sec^{-1}, selected so that each one satisfies $k_w + k_u = 4$. The difference $k_w - k_u$ of the two parameters contributes to the temporal dynamics of the moments. The total number of protein molecules was chosen to be $n^{tot} = 10$, initially all unmodified, that is, $N(0) = 10$.

(according to (6.3)) and its partial derivatives:

$$\frac{\partial f}{\partial x} = -k_1 (q_1 + q_2 - 2x) - k_2, , \qquad \frac{\partial^2 f}{\partial x^2} = 2k_1 .$$

3. Construct the diffusion coefficient $g_{ik}(x)$ according to (6.8) and the partial derivative $\frac{\partial^2 g_{ik}}{\partial x \partial x^T}$ for each pair of species:

$$g = v_1 + v_2 = k_1 (q_1 - x) (q_2 - x) + k_2 x, \qquad \frac{\partial^2 g}{\partial x^2} = 2k_1 .$$

4. Construct sums $\frac{\partial^2 f_i}{\partial x \partial x^T} : \langle \delta X \delta X^T \rangle$ of elementwise products between the

Hessian of f_i and covariance matrix for each species:

$$\frac{\partial^2 f}{\partial x^2} \langle \delta X^2 \rangle = 2k_1 \langle \delta X^2 \rangle .$$

5. Construct the scalar products $\frac{\partial f_i}{\partial x^T} \langle \delta X \delta X_k \rangle$ and $\langle \delta X_i \delta X^T \rangle \frac{\partial f_k}{\partial x}$ for each pair of species:

$$\frac{\partial f}{\partial x} \langle \delta X^2 \rangle = -\left[k_1 \left(q_1 + q_2 - 2\langle X \rangle \right) + k_2 \right] \langle \delta X^2 \rangle .$$

6. Construct sums $\frac{\partial^2 g_{ik}}{\partial x \partial x^T} : \langle \delta X \delta X^T \rangle$ of elementwise products between the Hessian of g_{ik} and covariance matrix for each pair of species:

$$\frac{\partial^2 g}{\partial x^2} \langle \delta X^2 \rangle = 2k_1 \langle \delta X^2 \rangle .$$

7. Insert the expressions obtained so far in (6.5) and (6.10) to obtain the 2MA equations

$$\frac{d \langle X \rangle}{dt} = k_1 \left(q_1 - \langle X \rangle \right) \left(q_2 - \langle X \rangle \right) - k_2 \langle X \rangle + k_1 \langle \delta X^2 \rangle ,$$

$$\frac{d \langle \delta X^2 \rangle}{dt} = 2\left[k_1 \left(\frac{1}{2\Omega} + 2\langle X \rangle - q_1 - q_2 \right) - k_2 \right] \langle \delta X^2 \rangle$$
$$+ \frac{1}{\Omega} \left[k_1 \left(q_1 - \langle X \rangle \right) \left(q_2 - \langle X \rangle \right) + k_2 \langle X \rangle \right] .$$

These 2MA equations can be solved numerically to compute the time-varying mean and variance of the product abundance for this example. The 2MA computed mean and the error bars determined by the mean±SD are plotted in Figure 6.3 together with the ensemble mean and the associated error bars, which were computed over 1000 realizations (SSA runs) of the stochastic process.

Example 6.5 (Lotka–Volterra model) For the reaction scheme (2.7) and the associated state transitions in Figure 5.24 of the Lotka–Volterra model, the 2MA equations are obtained by the same procedure outlined earlier.

1. Assign propensity $a_j(n)$ to each reaction channel:

$$a_1 = \hat{k}_1 n_A n_1, \quad a_2 = \hat{k}_2 n_1 n_2, \quad a_3 = \hat{k}_3 n_2 .$$

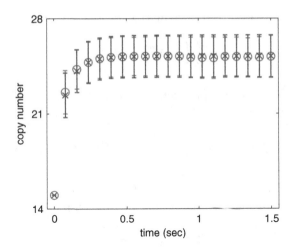

Figure 6.3 Mean and the error bars determined by the mean±SD of the product abundance for the heterodimerization reaction (2.6). The 2MA computed mean (*stars*) as well as the error bars match with the ensemble mean (*circles*) and the associated error bars, which were computed over 1000 realizations (SSA runs) of the stochastic process. The parameters and initial conditions are taken from Figure 5.5.

2. Construct the drift rate,

$$
A(n) = \begin{bmatrix} a_1 - a_2 \\ a_2 - a_3 \end{bmatrix} = \begin{bmatrix} \hat{k}_1 n_{\mathrm{A}} n_1 - \hat{k}_2 n_1 n_2 \\ \hat{k}_2 n_1 n_2 - \hat{k}_3 n_2 \end{bmatrix},
$$

(according to (5.29)) and its elementwise partial derivatives:

$$
\frac{\partial A}{\partial n^T} = \begin{bmatrix} \hat{k}_1 n_{\mathrm{A}} - k_2 \langle N_2 \rangle & -\hat{k}_2 \langle N_1 \rangle \\ \hat{k}_2 \langle N_2 \rangle & \hat{k}_2 \langle N_1 \rangle - \hat{k}_3 \end{bmatrix},
$$

$$
\frac{\partial^2 A_2}{\partial n \partial n^T} = \begin{bmatrix} 0 & \hat{k}_2 \\ \hat{k}_2 & 0 \end{bmatrix} = -\frac{\partial^2 A_1}{\partial n \partial n^T}.
$$

3. Construct the diffusion coefficient

$$
B = \begin{bmatrix} a_1 + a_2 & -a_2 \\ -a_2 & a_2 + a_3 \end{bmatrix} = \begin{bmatrix} \hat{k}_1 n_A n_1 + \hat{k}_2 n_1 n_2 & -\hat{k}_2 n_1 n_2 \\ -\hat{k}_2 n_1 n_2 & \hat{k}_2 n_1 n_2 + \hat{k}_3 n_2 \end{bmatrix},
$$

(according to (5.31)) and its elementwise partial derivatives:

$$
\frac{\partial^2 B_{11}}{\partial n \partial n^T} = \frac{\partial^2 B_{22}}{\partial n \partial n^T} = \begin{bmatrix} 0 & \hat{k}_2 \\ \hat{k}_2 & 0 \end{bmatrix} = -\frac{\partial^2 B_{12}}{\partial n \partial n^T} = -\frac{\partial^2 B_{21}}{\partial n \partial n^T}.
$$

4. Construct sums $\frac{\partial^2 A_i}{\partial n \partial n^T} : \langle \delta N \delta N^T \rangle$ of elementwise products between the Hessian of A_i and covariance matrix for each species:

$$
\frac{\partial^2 A_2}{\partial n \partial n^T} : \langle \delta N \delta N^T \rangle = -\frac{\partial^2 A_1}{\partial n \partial n^T} : \langle \delta N \delta N^T \rangle = 2\hat{k}_2 \langle \delta N_1 \delta N_2 \rangle.
$$

5. Construct the scalar products $\frac{\partial A_i}{\partial n^T} \langle \delta N \delta N_k \rangle$ and $\langle \delta N_i \delta N^T \rangle \frac{\partial A_k}{\partial n}$ for each pair of species:

$$
\frac{\partial A_1}{\partial n^T} \langle \delta N \delta N_1 \rangle = \left(\hat{k}_1 n_A - \hat{k}_2 \langle N_2 \rangle \right) \langle \delta N_1^2 \rangle - \hat{k}_2 \langle N_1 \rangle \langle \delta N_1 \delta N_2 \rangle,
$$

$$
\frac{\partial A_1}{\partial n^T} \langle \delta N \delta N_2 \rangle = \left(\hat{k}_1 n_A - \hat{k}_2 \langle N_2 \rangle \right) \langle \delta N_1 \delta N_2 \rangle - \hat{k}_2 \langle N_1 \rangle \langle \delta N_2^2 \rangle,
$$

$$
\langle \delta N_1 \delta N^T \rangle \frac{\partial A_2}{\partial n} = \hat{k}_2 \langle N_2 \rangle \langle \delta N_1^2 \rangle + \left(\hat{k}_2 \langle N_1 \rangle - \hat{k}_3 \right) \langle \delta N_1 \delta N_2 \rangle,
$$

$$
\frac{\partial A_2}{\partial n^T} \langle \delta N \delta N_2 \rangle = \hat{k}_2 \langle N_2 \rangle \langle \delta N_1 \delta N_2 \rangle + \left(\hat{k}_2 \langle N_1 \rangle - \hat{k}_3 \right) \langle \delta N_2^2 \rangle.
$$

6. Construct sums $\frac{\partial^2 B_{ik}}{\partial n \partial n^T} : \langle \delta N \delta N^T \rangle$ of elementwise products between the Hessian of B_{ik} and covariance matrix for each pair of species:

$$
\frac{\partial^2 B_{ik}}{\partial n \partial n^T} : \langle \delta N \delta N^T \rangle = \begin{cases} 2\hat{k}_2 \langle \delta N_1 \delta N_2 \rangle & \text{if } i = k, \\ -2\hat{k}_2 \langle \delta N_1 \delta N_2 \rangle & \text{if } i \neq k. \end{cases}
$$

7. Insert the expressions obtained so far in (6.4) and (6.9) to obtain the

2MA equations

$$\frac{d\langle N_1 \rangle}{dt} = \left(\hat{k}_1 n_A - k_2 \langle N_2 \rangle\right) \langle N_1 \rangle - \hat{k}_2 \langle \delta N_1 \delta N_2 \rangle,$$

$$\frac{d\langle N_2 \rangle}{dt} = \left(\hat{k}_2 \langle N_1 \rangle - \hat{k}_3\right) \langle N_2 \rangle + \hat{k}_2 \langle \delta N_1 \delta N_2 \rangle,$$

$$\frac{d\langle \delta N_1^2 \rangle}{dt} = 2\left(\hat{k}_1 n_A - \hat{k}_2 \langle N_2 \rangle\right) \langle \delta N_1^2 \rangle - (2\langle N_1 \rangle - 1)\hat{k}_2 \langle \delta N_1 \delta N_2 \rangle$$
$$+ \left(\hat{k}_1 n_A + \hat{k}_2 \langle N_2 \rangle\right) \langle N_1 \rangle,$$

$$\frac{d\langle \delta N_1 \delta N_2 \rangle}{dt} = \hat{k}_2 \langle N_2 \rangle \langle \delta N_1^2 \rangle - \hat{k}_2 \langle N_1 \rangle \langle \delta N_2^2 \rangle - \hat{k}_2 \langle N_1 \rangle \langle N_2 \rangle$$
$$+ \left(\hat{k}_1 n_A - \hat{k}_3 + (\langle N_1 \rangle - \langle N_2 \rangle - 1)\hat{k}_2\right) \langle \delta N_1 \delta N_2 \rangle,$$

$$\frac{d\langle \delta N_2^2 \rangle}{dt} = 2\left(\hat{k}_2 \langle N_1 \rangle - \hat{k}_3\right) \langle \delta N_2^2 \rangle + (2\langle N_2 \rangle + 1)\hat{k}_2 \langle \delta N_1 \delta N_2 \rangle$$
$$+ \left(\hat{k}_2 \langle N_1 \rangle + \hat{k}_3\right) \langle N_2 \rangle.$$

We now need to compute the numerical solution of these 2MA equations to obtain the time-varying mean and variance of the prey and predator populations for the Lotka–Volterra model. The 2MA computed mean and the error bars determined by the mean±SD are plotted in Figure 6.4 side by side with the ensemble mean and the associated error bars, which were computed over 1000 realizations (SSA runs) of the stochastic process.

Example 6.6 (Enzyme kinetic reaction) For the enzyme kinetic reaction scheme (2.8) with the associated state transitions in Figure (5.23), the 2MA equations are obtained by the same procedure outlined earlier, but replacing copy numbers by concentrations.

1. Assign kinetics rates $v_j(x)$ to each reaction channel:

$$v_1 = \left(x_E^{tot} - x_{ES}\right) k_1 x_S, \quad v_2 = k_2 x_{ES}, \quad v_3 = k_3 x_{ES}.$$

2. Construct the drift rate

$$f(x) = \begin{bmatrix} -v_1 + v_2 \\ \\ v_1 - v_2 - v_3 \end{bmatrix} = \begin{bmatrix} -\left(x_E^{tot} - x_{ES}\right) k_1 x_S + k_2 x_{ES} \\ \\ \left(x_E^{tot} - x_{ES}\right) k_1 x_S - (k_2 + k_3) x_{ES} \end{bmatrix},$$

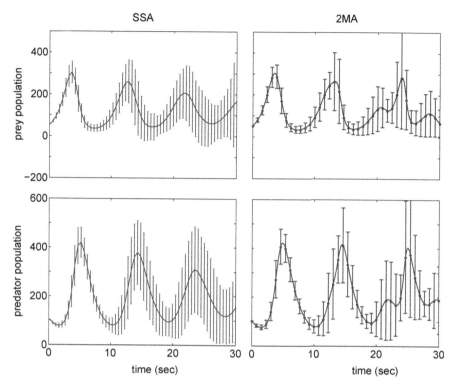

Figure 6.4 Mean and the error bars determined by the mean±SD of the species populations for the Lotka–Volterra model (2.7). *Left*: computed from 1000 realizations (SSA runs) of the stochastic process, *Right*: 2MA computed. The 2MA predictions match closely the corresponding SSA computed values. Note that the error bars have gone negative at some points, while the population must be nonnegative. That can be explained in two ways: (1) the probability distribution is not symmetric like a Gaussian (as assumed by error bars), and/or (2) extinction has occurred. The parameters and initial populations were taken from Figure 5.6.

(according to(6.3)) and its partial derivatives:

$$\frac{\partial f}{\partial x^T} = \begin{bmatrix} -\left(x_{\mathrm{E}}^{\mathrm{tot}} - \left\langle x_{\mathrm{ES}} \right\rangle\right) k_1 & k_1 \left\langle x_{\mathrm{S}} \right\rangle + k_2 \\ \left(x_{\mathrm{E}}^{\mathrm{tot}} - \left\langle x_{\mathrm{ES}} \right\rangle\right) k_1 & -k_1 \left\langle x_{\mathrm{S}} \right\rangle - k_2 - k_3 \end{bmatrix},$$

$$\frac{\partial^2 f_1}{\partial x \partial x^T} = \begin{bmatrix} 0 & k_1 \\ k_1 & 0 \end{bmatrix} = -\frac{\partial^2 f_2}{\partial x \partial x^T}.$$

3. Construct the diffusion coefficients

$$g_{11} = v_1 + v_2 = \left(x_{\mathrm{E}}^{\mathrm{tot}} - x_{\mathrm{ES}}\right) k_1 x_{\mathrm{S}} + k_2 x_{\mathrm{ES}},$$

$$g_{12} = g_{21} = -v_1 - v_2 = -\left(x_{\mathrm{E}}^{\mathrm{tot}} - x_{\mathrm{ES}}\right) k_1 x_{\mathrm{S}} - k_2 x_{\mathrm{ES}},$$

$$g_{22} = v_1 + v_2 + v_3 = \left(x_{\mathrm{E}}^{\mathrm{tot}} - x_{\mathrm{ES}}\right) k_1 x_{\mathrm{S}} + (k_2 + k_3) x_{\mathrm{ES}},$$

(according to (6.8)) and its elementwise partial derivatives:

$$\frac{\partial^2 g_{12}}{\partial x \partial x^T} = \frac{\partial^2 g_{21}}{\partial x \partial x^T} = \begin{bmatrix} 0 & k_1 \\ k_1 & 0 \end{bmatrix} = -\frac{\partial^2 g_{11}}{\partial x \partial x^T} = -\frac{\partial^2 g_{22}}{\partial x \partial x^T}.$$

4. Construct sums $\frac{\partial^2 f_i}{\partial x \partial x^T} : \langle \delta X \delta X^T \rangle$ of elementwise products between the Hessian of f_i and covariance matrix for each species:

$$\frac{\partial^2 f_1}{\partial x \partial x^T} : \langle \delta X \delta X^T \rangle = 2k_1 \langle \delta X_{\mathrm{S}} \delta X_{\mathrm{ES}} \rangle = -\frac{\partial^2 f_2}{\partial x \partial x^T} : \langle \delta X \delta X^T \rangle .$$

5. Construct the scalar products $\frac{\partial f_i}{\partial x^T} \langle \delta X \delta X_k \rangle$ and $\langle \delta X_i \delta X^T \rangle \frac{\partial f_k}{\partial x}$ for each pair of species:

$$\frac{\partial f_1}{\partial x^T} \langle \delta X \delta X_{\mathrm{S}} \rangle = -\left(x_{\mathrm{E}}^{\mathrm{tot}} - \langle X_{\mathrm{ES}} \rangle \right) k_1 \langle \delta X_{\mathrm{S}}^2 \rangle$$
$$+ (k_1 \langle X_{\mathrm{S}} \rangle + k_2) \langle \delta X_{\mathrm{S}} \delta X_{\mathrm{ES}} \rangle ,$$

$$\frac{\partial f_1}{\partial x^T} \langle \delta X \delta X_{\mathrm{ES}} \rangle = -\left(x_{\mathrm{E}}^{\mathrm{tot}} - \langle X_{\mathrm{ES}} \rangle \right) k_1 \langle \delta X_{\mathrm{S}} \delta X_{\mathrm{ES}} \rangle$$
$$+ (k_1 \langle X_{\mathrm{S}} \rangle + k_2) \langle \delta X_{\mathrm{ES}}^2 \rangle ,$$

$$\langle \delta X_{\mathrm{S}} \delta X^T \rangle \frac{\partial f_2}{\partial x} = \left(x_{\mathrm{E}}^{\mathrm{tot}} - \langle X_{\mathrm{ES}} \rangle \right) k_1 \langle \delta X_{\mathrm{S}}^2 \rangle$$
$$- (k_1 \langle X_{\mathrm{S}} \rangle + k_2 + k_3) \langle \delta X_{\mathrm{S}} \delta X_{\mathrm{ES}} \rangle ,$$

$$\frac{\partial f_2}{\partial x^T} \langle \delta X \delta X_{\mathrm{ES}} \rangle = \left(x_{\mathrm{E}}^{\mathrm{tot}} - \langle X_{\mathrm{ES}} \rangle \right) k_1 \langle \delta X_{\mathrm{S}} \delta X_{\mathrm{ES}} \rangle$$
$$- (k_1 \langle X_{\mathrm{S}} \rangle + k_2 + k_3) \langle \delta X_{\mathrm{ES}}^2 \rangle .$$

6. Construct sums $\frac{\partial^2 g_{ik}}{\partial x \partial x^T} : \langle \delta X \delta X^T \rangle$ of elementwise products between the

Hessian of g_{ik} and covariance matrix for each pair of species:

$$\frac{\partial^2 g_{ik}}{\partial x \partial x^T} : \langle \delta X \delta X^T \rangle = \begin{cases} -2k_1 \langle \delta X_S \delta X_{ES} \rangle & \text{if } i = k, \\ 2k_1 \langle \delta X_S \delta X_{ES} \rangle & \text{if } i \neq k. \end{cases}$$

7. Insert the expressions obtained so far in (6.5) and (6.10) to obtain the 2MA equations

$$\frac{d \langle X_S \rangle}{dt} = -\left(x_E^{tot} - \langle X_{ES} \rangle\right) k_1 \langle X_S \rangle + k_2 \langle X_{ES} \rangle + k_1 \langle \delta X_S \delta X_{ES} \rangle,$$

$$\frac{d \langle X_{ES} \rangle}{dt} = \left(x_E^{tot} - \langle X_{ES} \rangle\right) k_1 \langle X_S \rangle - (k_2 + k_3) \langle X_{ES} \rangle$$
$$- k_1 \langle \delta X_S \delta X_{ES} \rangle,$$

$$\frac{d \langle \delta X_S^2 \rangle}{dt} = -2 \left(x_E^{tot} - \langle X_{ES} \rangle\right) k_1 \langle \delta X_S^2 \rangle$$
$$+ \left(2k_1 \left\langle x_S \right\rangle + 2k_2 - \frac{k_1}{\Omega}\right) \langle \delta X_S \delta X_{ES} \rangle$$
$$+ \frac{1}{\Omega} \left[\left(x_E^{tot} - \langle X_{ES} \rangle\right) k_1 \langle X_S \rangle + k_2 \langle X_{ES} \rangle\right],$$

$$\frac{d \langle \delta X_S \delta X_{ES} \rangle}{dt} = \left(x_E^{tot} - \langle X_{ES} \rangle\right) k_1 \langle \delta X_S^2 \rangle + (k_1 \langle X_S \rangle + k_2) \langle \delta X_{ES}^2 \rangle$$
$$- \left[\left(\frac{1}{\Omega} + x_E^{tot} + \langle X_S \rangle - \langle X_{ES} \rangle\right) k_1 + k_2 + k_3\right] \langle \delta X_S \delta X_{ES} \rangle$$
$$- \frac{1}{\Omega} \left[\left(x_E^{tot} - \langle X_{ES} \rangle\right) k_1 \langle X_S \rangle + k_2 \langle X_{ES} \rangle\right],$$

$$\frac{d \langle \delta X_{ES}^2 \rangle}{dt} = -\left(2k_1 \langle X_S \rangle + k_2 + k_3\right) \langle \delta X_{ES}^2 \rangle$$
$$+ \left(2 \left(x_E^{tot} - \langle X_{ES} \rangle\right) - \frac{1}{\Omega}\right) k_1 \langle \delta X_S \delta X_{ES} \rangle$$
$$+ \frac{1}{\Omega} \left[\left(x_E^{tot} - \langle X_{ES} \rangle\right) k_1 \langle X_S \rangle + \left(k_2 + k_3\right) \langle X_{ES} \rangle\right].$$

The numerical solution of these 2MA equations provides us the time-varying mean and variance of the substrate and complex abundances. The 2MA computed mean and the error bars determined by the mean±SD are plotted in Figure 6.5 together with the ensemble mean and the associated error bars, which were computed over 1000 realizations (SSA runs) of the stochastic process.

Example 6.7 (Stochastic focusing) For the simplified branched reaction

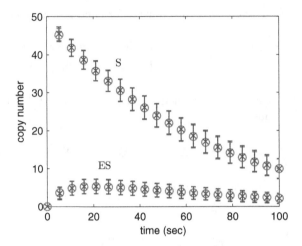

Figure 6.5 Mean and the error bars determined by the mean±SD of the substrate and complex abundances for the enzymatic reaction (2.8). The 2MA computed mean (*stars*) as well as the error bars match closely the ensemble mean (*circles*) and the associated error bars, which were computed over 1000 realizations (SSA runs) of the stochastic process. The parameters and initial conditions are taken from Figure 2.7.

scheme with the associated state transitions in Figure (5.21), the 2MA equations are obtained by the same procedure outlined earlier.

1. Assign propensities $a_j(n)$ to each reaction channel: (recall that $\Omega = 1$)

$$a_s^+ = k_s, \quad a_s^- = k_d n_S, \quad a_p^+ = \frac{k_i}{1 + \frac{n_S}{K}}, \quad a_p^- = n_P \, .$$

2. Construct the drift rate,

$$A(n) = \begin{bmatrix} a_s^+ - a_s^- \\ a_p^+ - a_p^- \end{bmatrix} = \begin{bmatrix} k_s - k_d n_S \\ \frac{k_i}{1 + \frac{n_S}{K}} - n_P \end{bmatrix},$$

(according to (5.29)) and its partial derivatives:

$$\frac{\partial A}{\partial n^T} = \begin{bmatrix} -k_d & 0 \\ \frac{-k_i/K}{\left(1 + \frac{n_S}{K}\right)^2} & -1 \end{bmatrix}, \quad \frac{\partial^2 A_1}{\partial n \partial n^T} = \begin{bmatrix} 0 & 0 \\ 0 & 0 \end{bmatrix}, \quad \frac{\partial^2 A_2}{\partial n \partial n^T} = \begin{bmatrix} \frac{2k_i/K^2}{\left(1 + \frac{n_S}{K}\right)^3} & 0 \\ 0 & 0 \end{bmatrix}.$$

3. Construct the diffusion coefficient,

$$
B(n) = \begin{bmatrix} a_s^+ + a_s^- & 0 \\ 0 & a_p^+ + a_p^- \end{bmatrix} = \begin{bmatrix} k_s + k_d n_S & 0 \\ 0 & \frac{k_i}{1 + \frac{n_S}{K}} + n_P \end{bmatrix},
$$

(according to (5.31)) and its elementwise partial derivatives:

$$
\frac{\partial^2 B_{11}}{\partial n \partial n^T} = \frac{\partial^2 B_{12}}{\partial n \partial n^T} = \frac{\partial^2 B_{21}}{\partial n \partial n^T} = \begin{bmatrix} 0 & 0 \\ 0 & 0 \end{bmatrix}, \quad \frac{\partial^2 B_{22}}{\partial n \partial n^T} = \begin{bmatrix} \frac{2k_i/K^2}{\left(1 + \frac{n_S}{K}\right)^3} & 0 \\ 0 & 0 \end{bmatrix}.
$$

4. Construct sums $\frac{\partial^2 A_i}{\partial n \partial n^T} : \langle \delta N \delta N^T \rangle$ of elementwise products between the Hessian of A_i and covariance matrix for each species:

$$
\frac{\partial^2 A_1}{\partial n \partial n^T} : \langle \delta N \delta N^T \rangle = 0, \quad \frac{\partial^2 A_2}{\partial n \partial n^T} : \langle \delta N \delta N^T \rangle = \frac{2k_i \langle \delta N_S^2 \rangle}{K^2 \left(1 + \frac{n_S}{K}\right)^3}.
$$

5. Construct the scalar products $\frac{\partial A_i}{\partial n^T} \langle \delta N_i \delta N_k \rangle$ and $\langle \delta N_i \delta N^T \rangle \frac{\partial A_k}{\partial n}$ for each pair of species:

$$
\frac{\partial A_1}{\partial n^T} \langle \delta N \delta N_S \rangle = -k_d \langle \delta N_S^2 \rangle,
$$

$$
\frac{\partial A_1}{\partial n^T} \langle \delta N \delta N_P \rangle = -k_d \langle \delta N_S \delta N_P \rangle,
$$

$$
\langle \delta N_S \delta N^T \rangle \frac{\partial A_2}{\partial n} = -\frac{k_i \langle \delta N_S^2 \rangle}{K \left(1 + \frac{\langle N_S \rangle}{K}\right)^2} - \langle \delta N_S \delta N_P \rangle,
$$

$$
\frac{\partial A_2}{\partial n^T} \langle \delta N \delta N_P \rangle = -\frac{k_i \langle \delta N_S \delta N_P \rangle}{K \left(1 + \frac{\langle N_S \rangle}{K}\right)^2} - \langle \delta N_P^2 \rangle.
$$

6. Construct sums $\frac{\partial^2 B_{ik}}{\partial n \partial n^T} : \langle \delta N \delta N^T \rangle$ of elementwise products between the Hessian of B_{ik} and covariance matrix for each pair of species:

$$
\frac{\partial^2 B_{ik}}{\partial n \partial n^T} : \langle \delta N \delta N^T \rangle = \begin{cases} \dfrac{2k_i \langle \delta N_S^2 \rangle}{K^2 \left(1 + \frac{n_S}{K}\right)^3} & \text{if } i = k = 2, \\ 0 & \text{otherwise.} \end{cases}
$$

7. Insert the expressions obtained so far in (6.4) and (6.9) to obtain the

2MA equations

$$\frac{d\langle N_S \rangle}{dt} = k_s - k_d \langle N_S \rangle$$

$$\frac{d\langle N_P \rangle}{dt} = -\langle N_P \rangle + \frac{k_i}{1 + \frac{\langle N_S \rangle}{K}} + \frac{k_i \langle \delta N_S^2 \rangle}{K^2 \left(1 + \frac{\langle N_S \rangle}{K}\right)^3},$$

$$\frac{d\langle \delta N_S^2 \rangle}{dt} = -2k_d \langle \delta N_S^2 \rangle + k_d \langle N_S \rangle + k_s,$$

$$\frac{d\langle \delta N_S \delta N_P \rangle}{dt} = -\frac{k_i \langle \delta N_S^2 \rangle}{K \left(1 + \frac{\langle N_S \rangle}{K}\right)^2} - (1 + k_d) \langle \delta N_S \delta N_P \rangle,$$

$$\frac{d\langle \delta N_P^2 \rangle}{dt} = -2\langle \delta N_P^2 \rangle + \frac{k_i \langle \delta N_S^2 \rangle}{K^2 \left(1 + \frac{\langle N_S \rangle}{K}\right)^3} - \frac{2k_i \langle \delta N_S \delta N_P \rangle}{K \left(1 + \frac{\langle N_S \rangle}{K}\right)^2}$$

$$+ \frac{k_i}{1 + \frac{\langle N_S \rangle}{K}} + \langle N_P \rangle .$$

Example 6.8 (Schlögl model) We saw in the last chapter that for the Schlögl reaction scheme (2.9) with state transitions in Figure 5.25, SSA-computed trajectories ultimately get distributed as separate clusters around the two fixed points. The mean copy number is located between the two clusters but does not represent any cluster. Therefore, the mean copy number is not an appropriate description for the Schlögl model, and in fact, for any multistable model. However, we will still attempt to construct 2MA equations for the Schlögl model to see the level of fluctuations relative to the mean. In a monostable system, one expects fluctuations to be proportional to the square root of the mean. In a bistable system, the fluctuations are comparable to the mean.

1. Assign propensity $a_j(n)$ to each reaction channel:

$$a_1 = (n-1)\hat{k}_1 n + \hat{k}_3, \quad a_2 = (n-2)(n-1)\hat{k}_2 n + k_4 n,$$

$$\text{where} \quad \hat{k}_1 = \frac{k_1 x_A}{\Omega}, \quad \hat{k}_2 = \frac{k_2}{\Omega^2}, \quad \hat{k}_3 = k_3 x_B \Omega .$$

2. Construct the drift rate,

$$A(n) = a_1 - a_2 = -\hat{k}_2 n^3 + \left(\hat{k}_1 + 3\hat{k}_2\right) n^2 - \left(k_4 + 2\hat{k}_2 + \hat{k}_1\right) n + \hat{k}_3,$$

(according to (5.29)) and its partial derivatives:

$$\frac{\partial A}{\partial n} = -3\hat{k}_2 \langle N \rangle^2 + 2\left(\hat{k}_1 + 3\hat{k}_2\right)\langle N \rangle - \hat{k}_1 - 2\hat{k}_2 - k_4,$$

$$\frac{\partial^2 A}{\partial n^2} = -6\hat{k}_2 \langle N \rangle + 2\hat{k}_1 + 6\hat{k}_2.$$

3. Construct the diffusion coefficient (according to (5.31)) and its partial derivative:

$$B = \hat{k}_2 n^3 + \left(\hat{k}_1 - 3\hat{k}_2\right)n^2 + \left(k_4 + 2\hat{k}_2 - \hat{k}_1\right)n + \hat{k}_3,$$

$$\frac{\partial^2 B}{\partial n^2} = 6\hat{k}_2 \langle N \rangle + 2\hat{k}_1 - 6\hat{k}_2.$$

4. Construct sums $\frac{\partial^2 A_i}{\partial n \partial n^T} : \langle \delta N \delta N^T \rangle$ of elementwise products between the Hessian of A_i and covariance matrix for each species:

$$\frac{\partial^2 A}{\partial n^2}\langle \delta N^2 \rangle = \left(-6\hat{k}_2 \langle N \rangle + 2\hat{k}_1 + 6\hat{k}_2\right)\langle \delta N^2 \rangle.$$

5. Construct the scalar (dot) products $\frac{\partial A_i}{\partial n^T}\langle \delta N \delta N_k \rangle$ and $\langle \delta N_i \delta N^T \rangle \frac{\partial A_k}{\partial n}$ for each pair of species:

$$\frac{\partial A}{\partial n}\langle \delta N^2 \rangle = \left[-3\hat{k}_2 \langle N \rangle^2 + 2\left(\hat{k}_1 + 3\hat{k}_2\right)\langle N \rangle - \hat{k}_1 - 2\hat{k}_2 - k_4\right]\langle \delta N^2 \rangle.$$

6. Construct sums $\frac{\partial^2 B_{ik}}{\partial n \partial n^T} : \langle \delta N \delta N^T \rangle$ of elementwise products between the Hessian of B_{ik} and covariance matrix for each pair of species:

$$\frac{\partial^2 B}{\partial n^2}\langle \delta N^2 \rangle = \left(6\hat{k}_2 \langle N \rangle + 2\hat{k}_1 - 6\hat{k}_2\right)\langle \delta N^2 \rangle.$$

7. Insert the expressions obtained so far in (6.4) and (6.9) to obtain the 2MA equations

$$\frac{d \langle N \rangle}{dt} = -\hat{k}_2 \langle N \rangle^3 + \left(\hat{k}_1 + 3\hat{k}_2\right)\langle N \rangle^2 - \left(k_4 + 2\hat{k}_2 + \hat{k}_1\right)\langle N \rangle + \hat{k}_3$$
$$+ \left(\hat{k}_1 + 3\hat{k}_2 - 3\hat{k}_2 \langle N \rangle\right)\langle \delta N^2 \rangle,$$

$$\frac{d \langle \delta N^2 \rangle}{dt} = \left[\left(15 \langle N \rangle - 6 \langle N \rangle^2 - 7\right)\hat{k}_2 + 3\hat{k}_1 - 2k_4\right]\langle \delta N^2 \rangle + \hat{k}_2 \langle N \rangle^3$$
$$+ \left(\hat{k}_1 - 3\hat{k}_2\right)\langle N \rangle^2 + \left(k_4 + 2\hat{k}_2 - \hat{k}_1\right)\langle N \rangle + \hat{k}_3.$$

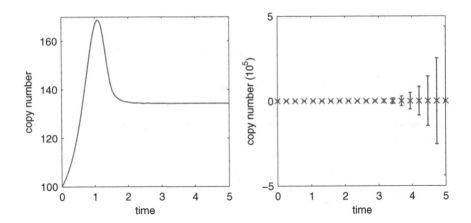

Figure 6.6 2MA predictions for the Schlögl reaction scheme (2.9). *Left*: Mean copy number, *Right*: mean together with the error bars determined by the mean±SD. The mean is overshadowed by the error bars. Since one would expect otherwise, the 2MA prediction suggests that further analysis is needed. The parameters and initial conditions are taken from Figure 1.7.

These 2MA equations can now be solved numerically. The results are plotted in Figure 6.6, wherein the mean copy number is shown alone on the left, and together with the error bars associated with mean±SD on the right. The mean is overshadowed by the error bars. Since one would expect otherwise, the 2MA prediction suggests that further analysis is needed.

The examples used above served as an illustration of the 2MA method. The next chapter investigates the 2MA approach for a practical example of a complex system with nonelementary reactions and relative concentration.

Problems

6.1. Consider the standard modification example with reaction scheme on the left and propensities on the right:

$$\left. \begin{array}{c} U \xrightarrow{k_w} W, \\ W \xrightarrow{k_u} U, \end{array} \right| \begin{array}{c} a_w(n) = k_w n, \\ a_u(n) = (n^{\text{tot}} - n)k_u, \end{array} \right\}$$

which is essentially a scalar system because of a conservation relation.

1. The following script constructs and solves the 2MA equations for this system.

```
ntot = 10; % total copy number
k = [2;2]; % rate constants
tmax = 5; % time scale in sec
y0 = [ntot;0]; % initial condition
S = [-1 1]; % Stoichiometry matrix
a = @ (n) [ k(1)*n ; k(2)*(ntot-n) ] ;
dydt = make2ma(S,a);
[t,y] = ode45(dydt, [0 tmax], y0);
```

The code calls the function make2ma (in the main text). Interpret the two columns of the output matrix y. Plot the two quantities against time.

2. To get an idea of the average behavior and fluctuations around it, plot the mean together with mean±SD.

6.2. The last exercise involved a scalar system of only one species. Consider the two-species Lotka–Volterra scheme and the associated propensities

$$
\left.
\begin{array}{ll}
X_1 + A \xrightarrow{\hat{k}_1} 2X_1, & a_1(n) = \hat{k}_1 n_A n_1, \\[2mm]
X_1 + X_2 \xrightarrow{\hat{k}_2} 2X_2, & a_2(n) = \hat{k}_2 n_1 n_2, \\[2mm]
X_2 \xrightarrow{\hat{k}_3} \varnothing, & a_3(n) = \hat{k}_3 n_2 .
\end{array}
\right\}
$$

The following script constructs and solves the 2MA equations for this system.

```
k = [1 .005 0.6];
tmax = 30; % time scale in sec
y0 = [50;100;zeros(3,1)]; % initial condition
S = [ 1  -1   0
      0   1  -1 ];
a = @ (n) [ k(1)*n(1) ; k(2)*n(1)*n(2) ; k(3)*n(2) ] ;
dydt = make2ma(S,a);
[t,y] = ode45(dydt, [0 tmax], y0);
```

Recover, for both species, the mean copy number and the corresponding standard deviation from the output matrix y.

6.3. Revisit the metabolite network in Exercise 2.6. In the follow-up exercises of the last chapter, you implemented computation of propensities and sample paths based on SSA and CLE. Use that information for tasks in this exercise.

1. Construct the 2MA equations. To keep only a few variables, you could use the reduced system (resulting from conservation relations) with abundances of X_2 and X_3 only.

2. Translate the 2MA equations to a Matlab function handle by calling the function `make2ma` with appropriate arguments.

3. Solve the the 2MA equations numerically and plot the mean together with mean±SD. Compare the 2MA predictions with results based on stochastic simulations.

6.4. Recall the repressilator in Exercise 2.7, for which you worked out the stoichiometry matrix S and the reaction propensity function $a(n)$ in the follow-up exercises in the last chapter. Now you have to revisit it from a 2MA perspective.

1. Construct the 2MA equations. For convenience, use a Hill's coefficient of $h = 2$.

2. Translate the 2MA equations to a Matlab function handle by calling the function `make2ma` with appropriate arguments.

3. Solve the the 2MA equations numerically and plot the mean abundance of proteins together with the corresponding mean±SD. Compare the 2MA predictions with results based on stochastic simulations.

Chapter 7

The 2MA Cell Cycle Model

In this chapter we take the Tyson–Novák model [108] for the fission yeast cell cycle as a case study [159]. This deterministic model is a practical example using nonelementary reactions and relative concentrations, the two central features of our extended 2MA approach. This will allow us to investigate the price of higher-order truncations by comparing the simulated cycle time statistics with experiments.

7.1 The 2MA Equations Revisited

In this chapter, we adopt a simplified notation for the relative concentration vector $X(t)$ with elements

$$X_i = \frac{N_i}{\Omega_i} = \frac{N_i}{C_i \Omega},$$

mean (relative) concentration vector $\mu(t)$ with elements

$$\mu_i = \langle X_i \rangle = \frac{\langle N_i \rangle}{C_i \Omega},$$

and the concentration covariance matrix $\sigma(t)$ with elements

$$\sigma_{ik} = \langle \delta X_i \delta X_k \rangle = \frac{\langle \delta N_i \delta N_k \rangle}{C_i C_k \Omega}$$

where, as obvious from the context, we leave out dependence on time. Recall that C_i are the componentwise characteristic concentrations used to normalize usual concentrations to obtain relative concentrations.

The two-moment equations (6.5) and (6.10), in the simplified notation, take the form

$$\frac{\mathrm{d}\mu_i}{\mathrm{d}t} = f_i(\mu) + \frac{1}{2}\frac{\partial^2 f_i}{\partial x \partial x^T} : \sigma, \tag{7.1}$$

$$\frac{\mathrm{d}\sigma_{ik}}{\mathrm{d}t} = \sum_l \left[\frac{\partial f_i}{\partial x_l}\sigma_{lk} + \sigma_{il}\frac{\partial f_k}{\partial x_l} \right] + \frac{1}{\Omega}\left[g_{ik}(\mu) + \frac{1}{2}\frac{\partial^2 g_{ik}}{\partial x \partial x^T} : \sigma \right], \tag{7.2}$$

where

$$f_i(x) = \frac{1}{C_i}\sum_{j=1}^{r} S_{ij}v_j(C \odot x),$$

$$\tag{7.3}$$

$$g_{ik}(x) = \frac{1}{C_i C_k}\sum_{j=1}^{r} S_{ij}S_{kj}v_j(C \odot x).$$

Recall that C is the s-vector of characteristic concentrations C_i, and the binary operation $\odot$ denotes the elementwise product of two arrays. The drift rate on the right in (7.1) has a deterministic part $f_i(\mu)$ and a stochastic part $\frac{1}{2}\frac{\partial^2 f_i}{\partial x \partial x^T} : \sigma$ determined by the dynamics of both the mean and (co)variance. This influence of the (co)variance implies that knowledge of fluctuations is important for a correct description of the mean. This also indicates an advantage of the stochastic framework over its deterministic counterpart: starting from the same assumptions and approximations, the stochastic framework allows us to describe the influence of fluctuations on the mean. This can be posed as the central phenomenological argument for stochastic modeling.

The scaling by Ω confirms the inverse relationship between the noise, as measured by (co)variance, and the system size. Note the influence of the mean on the (co)variance in (7.2).

Since the 2MA approach is based on the truncation of terms containing third and higher-order moments, any conclusion from the solution of 2MA must be drawn with care. Ideally, the 2MA should be complemented and checked with a reasonable number of SSA runs.

In [58, 62], the 2MA has been applied to biochemical systems, demonstrating quantitative and qualitative differences between the mean of the stochastic model and the solution of the deterministic model. The examples used in [58, 62] all assume elementary reactions (and hence propensities at most quadratic) and the usual interpretation of concentration as moles per unit volume. In the next section, we investigate the 2MA for complex systems with non-elementary and relative concentrations. The reason for our interest in nonelementary reactions is the frequent occurrence of rational propensities (reaction rates), e.g., Michaelis–Menten type and Hill type kinetic laws, in models in the systems biology literature (e.g., [157]).

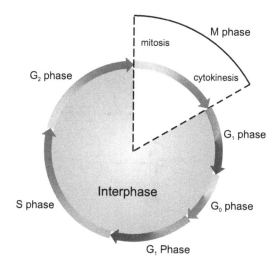

Figure 7.1
Phases of cell cycle regulation.
Figure adopted from [4].

7.2 Fission Yeast Cell Cycle Modeling

The growth and reproduction of organisms requires a precisely controlled sequence of events known as the cell cycle [4, 100]. On a coarse scale, the cell cycle is composed of four phases: the replication of DNA (S phase), the separation of DNA (mitosis, M-phase), and the intervening phases (gaps G1 and G2), which allow for preparation, regulation, and control of cell division. These phases are illustrated in Figure 7.1 for a generic cell cycle. The central molecular components of cell cycle control system have been identified [100, 110].

Cell cycle experiments show that cycle times (CTs) have different patterns for the wild type and for various mutants [145, 146]. For the wild type, the CTs have almost a constant value near 150 min ensured by a size control mechanism: mitosis happens only when the cell has reached a critical size. The double mutants (namely *wee1⁻ cdc25*Δ) exhibits quantized cycle times: the CTs get clustered into three different groups (with mean CTs of 90, 160, and 230 min). The proposed explanation for the quantized cycle times is a weakened positive feedback loop (due to wee1 and cdc25), which means that cells reset (more than once) back to G2 from early stages of mitosis by premature activation of a negative feedback loop [145, 147].

Many deterministic ODE models describing the cell cycle dynamics have been constructed [105, 107, 108, 156]. These models can explain many aspects of the cell cycle including the size control for both the wild type and mutants. Since deterministic models describe the behavior of a nonexistent 'average cell', neglecting the differences among cells in culture, they fail to

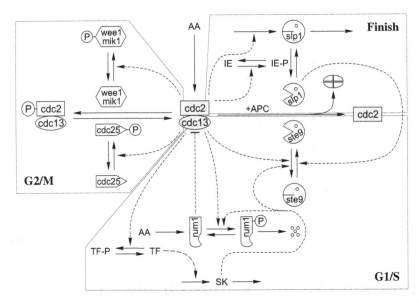

Figure 7.2 Regulation of fission yeast cell cycle. Figure reproduced from [108].

explain curious behaviors such as the quantized cycle times in double mutants. To account for such curiosities in experiments, two stochastic models were constructed by Sveiczer: The first model [145, 147] introduces (external) noise into the rate parameter of the protein Pyp3. The second model [148] introduces noise into two cell and nuclear sizes after division asymmetry. Full stochastic models that treat all the time-varying protein concentrations as random variables are reported in [144, 172]. They provide a reasonable explanation for the size control in wild type and the quantized CTs in the double-mutant type. Both models employ the Langevin approach and hence require many simulation runs to provide an ensemble for computing the mean and (co)variance. However, the simulation results of stochastic models in [144, 145, 147, 148, 172] represent one trajectory (for a large number of successive cycles) of the many possible in the ensemble from which the CT statistics (time averages) are computed. We will see that the time averages computed from the 2MA simulation are for the ensemble of all trajectories.

7.2.1 The Deterministic Model

We base our 2MA model on the Tyson–Novák model, a deterministic ODE model for the fission yeast cell cycle, developed by John Tyson and Béla Novák in [108]. As shown in Figure 7.2, the cell cycle control mechanism centers on the M-phase promoting factor (MPF), the active form of the heterodimer

Table 7.1 Proteins and fluxes.

Index i	Protein X_i	Production flux $f_i^+(x)$	Elimination flux $f_i^-(x)$
1	Cdc13$_T$	$k_1 M$	$(k_2' + k_2'' x_3 + k_2''' x_5)\, x_1$
2	preMPF	$(x_1 - x_2)\, k_{\text{wee}}$	$(k_{25} + k_2' + k_2'' x_3 + k_2''' x_5)\, x_2$
3	Ste9	$\dfrac{(k_3' + k_3'' x_5)(1 - x_3)}{J_3 + 1 - x_3}$	$\dfrac{(k_4' x_8 + k_4 x_{\text{mpf}}) x_3}{J_4 + x_3}$
4	Slp1$_T$	$k_5' + \dfrac{k_5'' x_{\text{mpf}}^4}{J_4^4 + x_{\text{mpf}}^4}$	$k_6 x_4$
5	Slp1	$k_7 \dfrac{(x_4 - x_5) x_6}{J_7 + x_4 - x_5}$	$k_6 x_5 + k_8 \dfrac{x_5}{J_8 + x_5}$
6	IEP	$k_9 \dfrac{(1 - x_6) x_{\text{mpf}}}{J_9 + 1 - x_6}$	$k_{10} \dfrac{x_6}{J_{10} + x_6}$
7	Rum1$_T$	k_{11}	$(k_{12} + k_{12}' x_8 + k_2'' x_{\text{mpf}})\, x_7$
8	SK	$k_{13} x_{\text{tf}}$	$k_{14} x_8$

Cdc13/Cdc2, and its antagonistic interactions with enemies (Ste9, Slp1, Rum1) and the positive feedback with its friend Cdc25. These interactions, among many others, define a sequence of checkpoints to control the timing of cell cycle phases. The result is MPF activity oscillation between low (G1-phase), intermediate (S- and G2-phases), and high (M-phase) levels that is required for the correct sequence of cell cycle events. For simplicity, it is assumed that the cell divides functionally when MPF drops below 0.1.

Table 7.1 lists the proteins whose concentrations x_i, together with MPF concentration, are treated as dynamic variables that evolve according to

$$\frac{\mathrm{d}x_i}{\mathrm{d}t} = f_i^+(x) - f_i^-(x). \tag{7.4}$$

Here $f_i^+(x)$ is the *production flux* and $f_i^-(x)$ is the *elimination flux* of the ith protein. Note that the summands in the fluxes $f_i^+(x)$ and $f_i^-(x)$ are rates of reactions, most of which are nonelementary (summarizing many elementary reactions into a single step). Quite a few of these reaction rates have rational expressions, which requires the extended 2MA approach developed in this book. The MPF concentration x_{mpf} can be obtained from the algebraic relation

$$x_{\text{mpf}} = \frac{(x_1 - x_2)(x_1 - x_{\text{trim}})}{x_1},\tag{7.5}$$

where

$$\frac{\mathrm{d}M}{\mathrm{d}t} = \rho M,$$

$$x_{\text{trim}} = \frac{2x_1 x_7}{\Sigma + \sqrt{\Sigma^2 - 4x_1 x_7}},$$

$$x_{\text{tf}} = G\left(k_{15} M, k'_{16}, k''_{16} x_{\text{mpf}}, J_{15}, J_{16}\right),$$

$$k_{\text{wee}} = k'_{\text{wee}} + (k''_{\text{wee}} - k'_{\text{wee}}) \, G\left(V_{\text{awee}}, V_{\text{iwee}} x_{\text{mpf}}, J_{\text{awee}}, J_{\text{iwee}}\right), \tag{7.6}$$

$$k_{25} = k'_{25} + (k''_{25} - k'_{25}) \, G\left(V_{\text{a25}} x_{\text{mpf}}, V_{\text{i25}}, J_{\text{a25}}, J_{\text{i25}}\right),$$

$$\Sigma = x_1 + x_7 + K_{\text{diss}},$$

$$G(a,b,c,d) = \frac{2ad}{b - a + bc + ad + \sqrt{(b - a + bc + ad)^2 - 4(b-a)ad}}.$$

Note that the cell mass M is assumed to grow exponentially with rate ρ, and the concentrations $(x_{\text{trim}}, x_{\text{tf}}, k_{\text{wee}}, k_{25})$ are assumed to be in a pseudosteady state to simplify the model. Note that we use a slightly different notation: ρ for mass growth rate (instead of μ), x_{trim} for the concentration of trimer (all associations between Cdc13/Cdc2 and Rum1), and x_{tf} for TF concentration. We have to emphasize that the concentrations used in this model are relative and dimensionless. When one concentration is divided by another, the proportion is the same as a proportion of two copy numbers. Hence, such a concentration should not be interpreted as a copy number per unit volume (as misinterpreted in [172]). The parameters used in the Tyson–Novák model [108] are listed in Table 7.2.

The deterministic ODE model describes the behavior of an *average cell*, neglecting the differences among cells in culture. Since the model allows at most two MPF resettings from early mitosis back to G2 (leading to alternating short and long cycles) for the double-mutant yeast cells as shown in Figure 7.3, it fails to explain the experimentally observed clusters of the CT-vs-BM plot and the trimodal distribution of CT [145–148].

7.2.2 Feasibility of Gillespie Simulations

Ideally, we should repeat many runs of Gillespie SSA and compute our desired moments from the ensemble of those runs. At present, there are two problems which this. The first problem is the requirement of elementary reactions for SSA. The elementary reactions underlying the deterministic model [108] are not known. Many elementary steps have been simplified to obtain that model.

Table 7.2 Parameter values for the Tyson–Novák model of the cell cycle control in the fission yeast (wild type) [108]. All the parameters have units min^{-1}, except the J's, which are dimensionless Michaelis constants, and K_{diss}, which is a dimensionless equilibrium constant for trimer dissociation. For the double-mutant type, one makes the following three changes: $k''_{\text{wee}} = 0.3$, $k'_{25} = k''_{25} = 0.02$.

$k_{15} = 0.03$, $k'_2 = 0.03$, $k''_2 = 1$, $k'''_2 = 0.1$, $k'_3 = 1$, $k''_3 = 10$, $J_3 = 0.01$,
$k'_4 = 2$, $k_4 = 35$, $J_4 = 0.01$, $k'_5 = 0.005$, $k''_5 = 0.3$, $k_6 = 0.1$, $J_5 = 0.3$,
$k_7 = 1$, $k_8 = 0.25$, $J_7 = J_8 = 0.001$, $J_8 = 0.001$, $k_9 = 0.1$, $k_{10} = 0.04$,
$J_9 = 0.01$, $J_{10} = 0.01$, $k_{11} = 0.1$, $k_{12} = 0.01$,
$k'_{12} = 1$, $k''_{12} = 3$, $K_{\text{diss}} = 0.001$,
$k_{13} = 0.1$, $k_{14} = 0.1$, $k_{15} = 1.5$, $k'_{16} = 1$, $k''_{16} = 2$, $J_{15} = 0.01$, $J_{16} = 0.01$,
$V_{\text{awee}} = 0.25$, $V_{\text{iwee}} = 1$, $J_{\text{awee}} = 0.01$, $J_{\text{iwee}} = 0.01$, $V_{\text{a}25} = 1$, $V_{\text{i}25} = 0.25$,
$J_{\text{a}25} = 0.01$, $J_{\text{i}25} = 0.01$, $k'_{\text{wee}} = 0.15$, $k''_{\text{wee}} = 1.3$,
$k'_{25} = 0.05$, $k''_{25} = 5$, $\rho = 0.005$

Trying to perform an SSA on nonelementary reactions will lose the discrete event character of the SSA. The second problem arises from the fact that the SSA requires copy numbers, which in turn requires knowledge of measured concentrations. All protein concentrations in the model are expressed in arbitrary units (a.u.), because the actual concentrations of most regulatory proteins in the cell are not known [26]. Tyson and Sveiczer[1] define relative concentration x_i of the ith protein as $x_i = n_i/\Omega_i$, where $\Omega_i = C_i N_A V$. Here C_i is an unknown characteristic concentration of the ith component. The idea is to have a common scale for all the relative concentrations so that a comparison between changes in any two of them is free of their individual (different) scales. Although one would like to vary C_i, this is computationally intensive. This problem is not so serious for the continuous approximations such as CLE, LNA, and the 2MA which are all ODEs and can be numerically solved. Using Matlab R2009a on a quad-core 2.66 GHz CPU took longer than 10 hours to complete one SSA run of 465 cycles. According to a recently published report [3], to compare the stochastic results with the average behavior, the simulation must be run thousands of times, for which the cited authors had to use a parallel supercomputer. The main focus of the present chapter is the analytical 2MA.

---

[1] Personal communication.

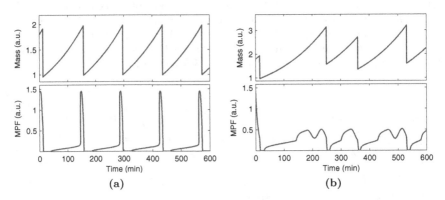

Figure 7.3 Time course according to the Tyson–Novák model: (a) the wild-type cells have cycles of nearly the same duration, (b) the double-mutant cells have alternating short (no MPF resetting) and long cycles (one MPF resetting).

7.2.3 The Langevin Approach

In [172] a stochastic model is proposed that replaces the ODE model (7.4) with a set of chemical Langevin equations (CLEs)

$$\frac{\mathrm{d}x_i}{\mathrm{d}t} = f_i^+(x) - f_i^-(x) + \frac{1}{\Omega}\left[\sqrt{f_i^+(x)}\Gamma_i^+(t) - \sqrt{f_i^-(x)}\Gamma_i^-(t)\right],$$

which uses the Langevin noise terms: white noises Γ_i^+ and Γ_i^- scaled by $\sqrt{f_i^+(x)}$ and $\sqrt{f_i^-(x)}$ to represent the internal noise. The system parameter Ω has been described as the volume by the author. As we discussed before, the concentrations are relative levels with possibly different componentwise scaling parameters. This means that concentrations are not the same as copy numbers per unit volume.

Another stochastic model employing the Langevin approach is reported in [144]. It approximates the squared noise amplitudes by linear functions:

$$\frac{\mathrm{d}x_i}{\mathrm{d}t} = f_i(x(t)) + \sqrt{2d_i x_i}\Gamma_i(t),$$

where d_i is a constant. The model dynamics $f(x)$ are missing in the noise term because the author wanted to represent both the internal and external noise by the second term on the right.

7.2.4 The 2MA Model

For the cell cycle model, the drift rate $f(x)$ and the diffusion rate g, defined in (7.3), have elements

$$f_i(x) = f_i^+(x) - f_i^-(x), \quad g_{ik}(x) = \begin{cases} f_i^+(x) + f_i^-(x) & \text{if } i = k, \\ 0 & \text{if } i \neq k. \end{cases}$$

The off-diagonal elements of g are zero because each reaction changes only one component, so that $S_{ij}S_{kj} = 0$ for $i \neq k$. Once these quantities are known, it follows from (7.1) and (7.2) that the set of ODEs

$$\frac{d\mu_i}{dt} = f_i(\mu) + \frac{1}{2}\frac{\partial^2 f_i}{\partial x \partial x^T} : \sigma, \tag{7.7}$$

$$\frac{d\sigma_{ii}}{dt} = 2\sum_l \frac{\partial f_i}{\partial x_l}\sigma_{li} + \frac{1}{\Omega_i}\left[g_{ii}(\mu) + \frac{1}{2}\frac{\partial^2 g_{ii}}{\partial x \partial x^T} : \sigma\right], \tag{7.8}$$

$$\frac{d\sigma_{ik}}{dt} = \sum_l \left[\frac{\partial f_i}{\partial x_l}\sigma_{lk} + \sigma_{il}\frac{\partial f_k}{\partial x_l}\right], \quad i \neq k, \tag{7.9}$$

approximates (correctly to the second-order moments) the evolution of componentwise concentration mean and covariance. See Tables 7.3–7.5 for the respective expressions of the Jacobian $\partial f/\partial x^T$, the second-order term $\frac{1}{2}\partial^2 f_i/\partial x \partial x^T : \sigma$ in the Taylor expansion of f_i, and the second-order term $\frac{1}{2}\partial^2 g_{ii}/\partial x \partial x^T : \sigma$ in the Taylor expansion of g_{ii} in (7.8).

Having at hand the moments involving the eight dynamic variables x_1 to x_8, the mean MPF concentration can also be approximated. Toward that end, we start with the MPF concentration

$$x_{\text{mpf}} = (x_1 - x_2)\left(1 - \frac{x_{\text{trim}}}{x_1}\right) = x_1 - x_2 - x_{\text{trim}} + x_{\text{trim}}\frac{x_2}{x_1}.$$

The ratio x_2/x_1 can be expanded around the mean,

$$\frac{x_2}{x_1} = \frac{1}{\mu_1}\frac{x_2}{1 + \frac{(x_1-\mu_1)}{\mu_1}} = \frac{1}{\mu_1}\left[x_2 - \frac{(x_1 - \mu_1)x_2}{\mu_1} + \frac{(x_1 - \mu_1)^2 x_2}{\mu_1^2} + \cdots\right].$$

Table 7.3 Rows of the Jacobian of the drift rate for the 2MA cell cycle model.

i	$\frac{\partial f_i}{\partial x^T}$
1	$\left[-k_2' - k_2''\mu_3 - k_2'''\mu_5, 0, -k_2''\mu_1, 0, -k_2'''\mu_1, 0, 0, 0\right]$
2	$\left[k_{\text{wee}}, -k_{\text{wee}} - k_{25} - k_2' - k_2''\mu_3 - k_2'''\mu_5, -k_2''\mu_2, 0, -k_2'''\mu_2, 0, 0, 0\right]$
3	$\left[0, 0, -\frac{(k_4'\mu_8 + k_4\mu_{\text{mpf}})J_4}{(J_4+\mu_3)^2} - \frac{(k_3'+k_3''\mu_5)J_3}{(J_3+1-\mu_3)^2}, 0, \frac{(1-\mu_3)k_3''}{J_3+1-\mu_3}, 0, 0, -\frac{k_4'\mu_3}{J_4+\mu_3}\right]$
4	$[0, 0, 0, -k_6, 0, 0, 0, 0]$
5	$\left[0, 0, 0, \frac{k_7 J_7\mu_6}{(J_7+\mu_4-\mu_5)^2}, -k_6 - \frac{k_7 J_7\mu_6}{(J_7+\mu_4-\mu_5)^2} - \frac{k_8 J_8}{(J_8+\mu_5)^2}, \frac{(\mu_4-\mu_5)k_7}{J_7+x_4-\mu_5}, 0, 0\right]$
6	$\left[0, 0, 0, 0, 0, -\frac{k_9 x_{\text{mpf}} J_9}{(J_9+1-\mu_6)^2} - \frac{k_{10} J_{10}}{(J_{10}+\mu_6)^2}, 0, 0\right]$
7	$\left[0, 0, 0, 0, 0, 0, -k_{12} - k_{12}'\mu_8 - k_2''\mu_{\text{mpf}}, -k_{12}'\mu_7\right]$
8	$\left[0, 0, 0, 0, 0, 0, 0, -k_{14}\right]$

Taking the expectation on both sides yields

$$\left\langle \frac{X_2}{X_1} \right\rangle = \frac{1}{\mu_1}\left\langle \frac{X_2}{1 + \frac{(X_1-\mu_1)}{\mu_1}} \right\rangle$$

$$= \frac{1}{\mu_1}\left\langle X_2 - \frac{(X_1-\mu_1)X_2}{\mu_1} + \frac{(X_1-\mu_1)^2 X_2}{\mu_1^2} + \cdots \right\rangle$$

$$= \frac{1}{\mu_1}\left[\mu_2 - \frac{\sigma_{12}}{\mu_1} + \frac{\mu_2\sigma_{11}}{\mu_1^2}\right].$$

Finally, with the understanding that x_{trim} is in pseudosteady state, the mean MPF concentration follows from the expectation of x_{mpf} to be

$$\mu_{\text{mpf}} = \mu_1 - \mu_2 - x_{\text{trim}} + \frac{x_{\text{trim}}}{\mu_1}\left[\left(1 + \frac{\sigma_{11}}{\mu_1^2}\right)\mu_2 - \frac{\sigma_{12}}{\mu_1}\right]. \qquad (7.10)$$

This expression for the average MPF activity demonstrates the influence of (co)variance on the mean, as emphasized here. We see the dependence of mean MPF concentration μ_{mpf} on the variance σ_{11} and covariance σ_{12} in

Table 7.4 The second-order term in the Taylor expansion of f_i around the mean.

i	$\frac{1}{2}\frac{\partial^2 f_i}{\partial x \partial x^T} : \sigma$
1	$-k_2''\sigma_{13} - k_2'''\sigma_{15}$
2	$-k_2''\sigma_{23} - k_2'''\sigma_{25}$
3	$\left[\dfrac{(k_4'\mu_8 + k_4\mu_{\text{mpf}})J_4}{(J_4+\mu_3)^3} - \dfrac{(k_3' + k_3''\mu_5)J_3}{(J_3+1-\mu_3)^3}\right]\sigma_{33} - \dfrac{k_3''J_3\sigma_{35}}{(J_3+1-\mu_3)^2} - \dfrac{k_4'J_4\sigma_{38}}{(J_4+\mu_3)^2}$
4	0
5	$\dfrac{k_7 J_7\mu_6(2\sigma_{45} - \sigma_{44} - \sigma_{55})}{(J_7+\mu_4-\mu_5)^3} + \dfrac{k_7 J_7(\sigma_{46} - \sigma_{56})}{(J_7+\mu_4-\mu_5)^2} + \dfrac{k_8 J_8}{(J_8+\mu_5)^3}\sigma_{55}$
6	$\left[\dfrac{k_{10} J_{10}}{(J_{10}+\mu_6)^3} - \dfrac{k_9\mu_{\text{mpf}}J_9}{(J_9+1-\mu_6)^3}\right]\sigma_{66}$
7	$-k_{12}'\sigma_{78}$
8	0

addition to the means μ_1, μ_2, and x_{trim}.

7.2.5 Simulations of the 2MA Model

The system of ODEs (7.7)–(7.9) was solved numerically by the Matlab solver ode15s [96]. The solution was then combined with algebraic relations (7.10). For parameter values, see Table 7.2. The system-size parameter is chosen to be $\Omega = 5000$. Since information about the individual characteristic concentrations C_i used in the definition of concentrations is not available, we have used $C_i = 1$ for all i. This value has also been used in [172], although there is no clear justification. The system size here should not be interpreted as volume, because that would imply knowledge of the scale of characteristic concentrations. Note, however, that the 2MA approach developed here will work for any combination of $\{C_i\}$. The time courses of mass and MPF activity are plotted in Figure 7.4a for the wild type and in Figure 7.4b for the double-mutant type. For the wild type, the 2MA-predicted mean trajectories do not differ considerably from the corresponding deterministic trajectories. Both show nearly a constant CT around 150 min. Thus internal fluctuations do not seem to have a major influence for the wild-type cells.

For the double-mutant cells, the difference between the 2MA and de-

Table 7.5 The second-order term in the Taylor expansion of g_{ii} around the mean.

i	$\frac{1}{2}\frac{\partial^2 g_{ii}}{\partial x \partial x^T} : \sigma$
1	$k_2'' \sigma_{13} + k_2''' \sigma_{15}$
2	$k_2'' \sigma_{23} + k_2''' \sigma_{25}$
3	$-\left[\dfrac{(k_4'\mu_8 + k_4\mu_{\mathrm{mpf}})J_4}{(J_4+\mu_3)^3} + \dfrac{(k_3'+k_3''\mu_5)J_3}{(J_3+1-\mu_3)^3}\right]\sigma_{33} - \dfrac{k_3'' J_3 \sigma_{35}}{(J_3+1-\mu_3)^2} + \dfrac{k_4' J_4 \sigma_{38}}{(J_4+\mu_3)^2}$
4	0
5	$\dfrac{k_7 J_7 \mu_6 (2\sigma_{45} - \sigma_{44} - \sigma_{55})}{(J_7+\mu_4-\mu_5)^3} + \dfrac{k_7 J_7 (\sigma_{46} - \sigma_{56})}{(J_7+\mu_4-\mu_5)^2} - \dfrac{k_8 J_8}{(J_8+\mu_5)^3}\sigma_{55}$
6	$-\left[\dfrac{k_{10} J_{10}}{(J_{10}+\mu_6)^3} + \dfrac{k_9 \mu_{\mathrm{mpf}} J_9}{(J_9+1-\mu_6)^3}\right]\sigma_{66}$
7	$k_{12}' \sigma_{78}$
8	0

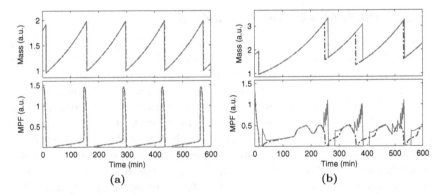

Figure 7.4 Time courses of mass and MPF activity: (a) for the wild type, (b) for the double-mutant type. The 2MA-predicted mean trajectories are plotted as solid lines and the corresponding deterministic trajectories as dashed lines. The difference between the two predictions is negligible for the wild type, but significant for the double-mutant type. The figure first appeared in our earlier work [159].

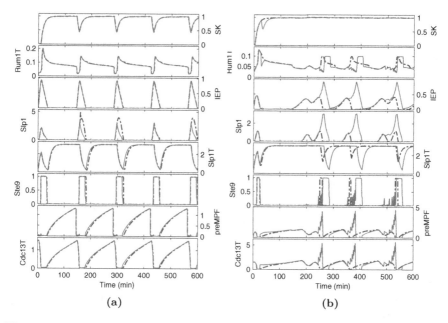

Figure 7.5 Time courses of protein concentrations: (a) for the wild type, (b) for the double-mutant type. The 2MA-predicted mean trajectories are plotted as solid lines and the corresponding deterministic trajectories as dashed lines.

terministic predictions is significant. The deterministic model (7.4) predicts alternating short cycles and long cycles because cells born at the larger size have shorter cycles, and smaller newborns have longer cycles [108]. This strict alternation due to size control is not observed in experiments: cells of the same mass may have short or long cycles (excluding very large cells that always have the shortest CT) [146, 147]. This weak size control is reproduced by the 2MA simulations: the multiple resettings of MPF to G2, induced by the internal noise, result in longer CTs (thus accounting for the 230-min cycles observed experimentally). Such MPF resettings have been proposed in [145, 147] to explain quantized CTs. No such resetting is demonstrated by the deterministic model.

Figure 7.5 additionally shows time courses of slp1, Ste9, and Rum1$_T$. For the wild type, the difference in the Rum1T concentrations near the G2/M transition has no significant effect on the MPF activity, because Rum1T tries to inhibit MPF in G2-phase. For the double-mutant type, the oscillatory behavior of Ste9 and Slp1 may have resulted in the oscillatory behavior of the MPF near the G2/M transition, which in turn delays the mitosis by a

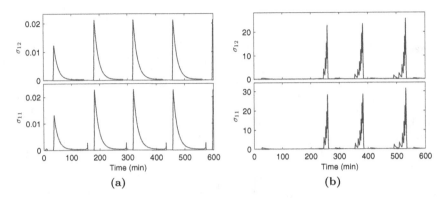

Figure 7.6 Variance σ_{11} (of Cdc13$_T$) and covariance σ_{12} (between Cdc13$_T$ and preMPF): (a) for the wild type, (b) for the double-mutant type. The figure first appeared in our earlier work [159].

noticeable period.

Note that the mean $\mu(t)$ of the 2MA describes the average of an ensemble of cells. Yet the MPF resettings observed in Figure (7.4b), near the G2/M transition, introduce the required variability that explains the clustering of the cycle time observed in experiments. This is in contrast to the alternative stochastic approaches in [144, 145, 147, 148, 172] that use one sample trajectory rather than the ensemble average.

How do we explain this significant effect of noise for the double mutants on the one hand and its negligible effect for the wild type on the other hand? If we look at expression (7.10), we see the influence of the variance σ_{11} (of Cdc13$_T$) and covariance σ_{12} (between Cdc13$_T$ and preMPF) on the mean MPF concentration μ_{mpf}. The two (co)variances are plotted in Figure 7.6a for the wild type and in Figure 7.6b for the double-mutant type. It is clear that the two (co)variances have very small peaks for the wild type compared to the large peaks for the double-mutant type. Note that the larger peaks in Figure 7.6b are located at the same time points where the MPF activity exhibits oscillations and hence multiple resettings to G2. This suggests that the oscillatory behavior of MPF near the G2/M transition is due to the influence of the oscillatory (co)variances. This coupling between the mean and (co)variance is not captured by the deterministic model.

The rapid MPF oscillations near entry to mitosis warrants further discussion. The authors of the Tyson–Novák model [108] show that G2/M transition is a point of bifurcation (between low MPF in G2 and high MPF in mitosis), considering the cell mass M as a parameter. That is exactly where the 2MA model predicts these rapid MPF oscillations. This suggests

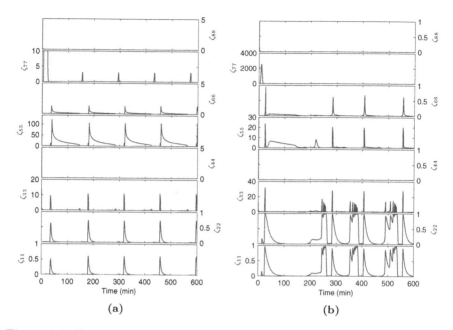

Figure 7.7 Noise-to-signal ratio (NSR): (a) for the wild type, (b) for the double-mutant type.

a noise induced change in the stability of the system near the critical point (G2/M transition) where a steady limit cycle (in G2) is changed into a chaotic unstable oscillation [70].

To allow for comparison between componentwise variances, the variance is usually normalized by the squared mean to give a dimensionless ratio, which also removes the dependence of the variance on the scale of the mean. The normalized variance,

$$\zeta_{ii} = \frac{\sigma_{ii}}{\mu_i^2},$$

is the *noise-to-signal ratio* (NSR). The NSR as a measure of noise is usually preferred because of being dimensionless and allowing for additivity of noise and the use of indirectly measured concentrations [58]. See [117] for different measures of noise and their merits.

The componentwise NSR is plotted in Figure 7.7. We note that the NSR for the double-mutant type has irregular oscillations compared to the almost-periodic oscillations for the wild type. This may be one of the reasons behind the significant difference in the evolution mean compared to the deterministic evolution for the double-mutant type. The pairwise variation

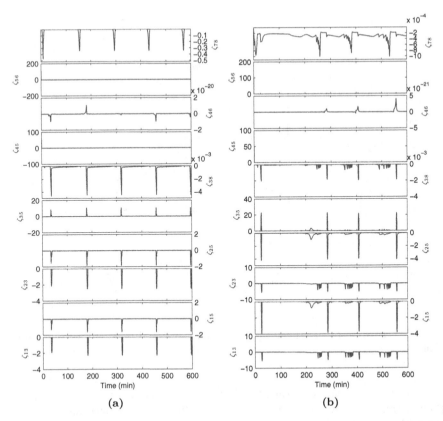

Figure 7.8 Cross noise-to-signal ratio (xNSR) of selected component pairs: (a) for the wild type, (b) for the double-mutant type.

between components is better described by the *cross* noise-to-signal ratio (xNSR)

$$\zeta_{ik} = \frac{\sigma_{ik}}{\mu_i^2}.$$

The xNSR for selected component pairs appearing in the expression for the stochastic part of the total drift (see Table 7.4) is plotted in Figure 7.8. The oscillatory behavior in both plots suggests that the off-diagonal elements of the covariance matrix may have influenced the mean and had a mutual influence on each other. At these points the system is sensitive to noise. Capturing these phenomena is of particular importance if one considers cells in their context (e.g., tissue), where cell–cell variations form the basis for functional mechanisms at higher levels of cellular organization.

Table 7.6 Cycle time statistics over 465 successive cell cycles of the double-mutant cells, predicted by the 2MA model, compared with experimental data; see [146, Table 1].

Case	μ_{CT}	σ_{CT}	CV_{CT}	μ_{DM}	σ_{DM}	CV_{DM}	μ_{BM}	σ_{BM}
(1)	131	47	0.358	2.22	0.45	0.203	1.21	0.24
(2)	138.8	12.4	0.09	3.18	0.101	0.0319	1.59	0.0575
(3)	138.8	17.6	0.127	3.25	0.178	0.055	1.623	0.0934
(4)	138.8	23.9	0.172	3.32	0.231	0.0697	1.657	0.12

(1) experimental data, (2) $\Omega = 5000$, (3) $\Omega = 5200$, (4) $\Omega = 5300$.

It has to be realized that the above proposition requires validation, since the 2MA approach ignores third and higher-order moments. We cannot know whether that truncation is responsible for the oscillations in Figures 7.4 and 7.6, unless compared with a few sample trajectories simulated by the SSA. However, as discussed before, the SSA cannot be performed (at present) for the model under consideration. Therefore we need to compare the 2MA predictions for the double-mutant cells with experimental data. Toward that end, values of cycle time (CT), birth mass (BM), and division mass (DM) were computed for 465 successive cycles of double-mutant cells. Figure 7.9 shows the CT-vs-BM plot and the CT distribution for three different values $\{5000, 5200, 5300\}$ of system size Ω.

To make this figure comparable with experimental data from [145, 146], we assume that 1 unit of mass corresponds to 8.2 μm cell length [147]. We can see the missing size control (CT clusters), in qualitative agreement with experimentally observed clusters (see [146, Figure 6] and [145, Figure 5] for a comparison). There are more than four clusters, which may have arisen from the truncated higher-order moments. The extreme value of CT higher than 230 min suggests more than two MPF resettings. Furthermore, more than three modes in the CT distribution may have arisen from the truncated higher-order moments. Table 7.6 compares the 2MA-computed statistics for the double-mutant cells with data from [146, Table 1]. Columns 2–4 tabulate, for CT, the mean μ_{CT}, the standard deviation σ_{CT}, and the coefficient of variation CV_{CT}, respectively. The other columns tabulate similar quantities for the division mass (DM) and birth mass (BM). We see that only the mean CT is in agreement with the experimental data. The mean values for both BM

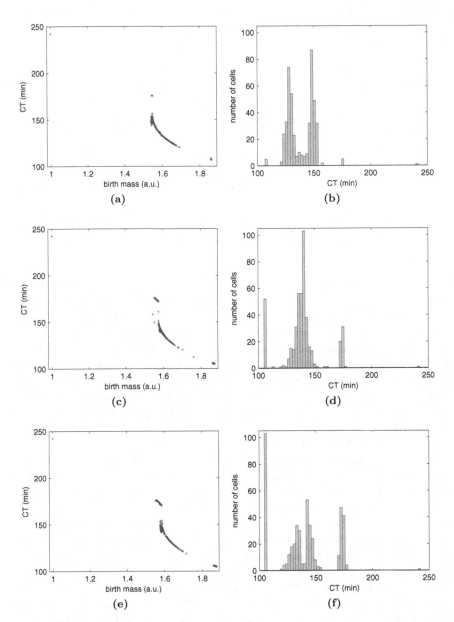

Figure 7.9 Cycle time behavior over 465 successive cycles of the double-mutant cells, predicted by the 2MA model. (a, c, e): CT vs BM, (b, d, f): CT distribution, (a, b): $\Omega = 5000$, (c,d): $\Omega = 5200$, (e,f): $\Omega = 5300$. The plots are in qualitative agreement with experiments; see Figure 6 in [146] and Figure 5 in [145] for a comparison. The figure first appeared in our earlier work [159].

and DM are larger than the corresponding experimental values. The other statistics are much smaller than the corresponding experimental values. This table and the above plots suggest that the 2MA should be used with caution. However, another aspect of the cell cycle model deserves attention here. The way the relative protein concentrations have been defined implies unknown values of the scaling parameters $\{\Omega_i\}$. Since $\Omega_i = C_i N_A V$, knowing the volume V does not solve the problem: the characteristic concentrations $\{C_i\}$ are still unknown. Our simulations have chosen typical values $\Omega = \{5000, 5200, 5300\}$. The corresponding three pairs of plots in Figure 7.9 and rows in Table 7.6 demonstrate a dependence of the results on a suitable system size. There is no way to confirm these values. The scaling parameters could be regulated in a wider range in order to improve the accuracy of our simulation, motivating future work for us. The conclusion is that the quantitative disagreement of the 2MA predictions can be attributed to two factors: (1) the truncated higher-order moments during the derivation of the 2MA, and (2) the unknown values of the scaling parameters.

7.2.6 Concluding Remarks

We investigated the applicability of the 2MA approach to the well established fission yeast cell cycle model. The simulations of the 2MA model show oscillatory behavior near the G2/M transition, which is significantly different from the simulations of the deterministic ODE model. One notable aspect of the analytical model is that although it describes the average of an ensemble, it reproduces enough variability among cycles to reproduce the curious quantized cycle times observed in experiments on double mutants.

Chapter 8

Hybrid Markov Processes

Our focus so far has been on continuous-time, discrete-state-space Markov jump processes. These stochastic processes are conceptually suitable to biochemical reaction networks because the reaction events occur randomly on a continuous time scale and bring about discrete changes in the species abundances. We have also seen how a jump process can be approximated by a diffusion process, which is a continuous process, using approaches including the chemical Langevin equation (and the associated Fokker–Planck equation) and the system size expansion. Apart from the two extremes is an intermediate possibility, a hybrid process that is essentially a diffusion process but with occasional jumps. More appropriate terms for such hybrid processes in the context of systems biology are "switch plus diffusion" and "diffusion process with Markovian switching" [72, 134]. This happens in two-time-scale systems wherein events occurring on the slower time scale contribute a jump process (Markovian switching) superimposed on a diffusion process arising from events occurring on the faster time scale. A sample path of such a hybrid process is a continuously varying time course with occasional jumps, as illustrated in Figure 8.1.

Yet another modeling regime keeps the switching (jumps) and drift but ignores diffusion and appears in the systems biology literature under the names "switch plus ODE" [72, 115] and "piecewise-deterministic Markov processes" (PDMP) [175]. A sample path of such a hybrid process is illustrated in Figure 8.1, where occasional jumps (occurring at a slower time scale) ride on top of a smooth drift process.

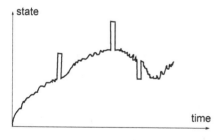

Figure 8.1
A sample path of a general hybrid Markov process. Occasional jumps (occurring at a slower time scale) ride on top of a smooth process with diffusion and drift.

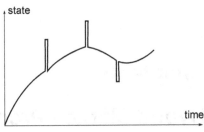

Figure 8.2
A sample path of a hybrid Markov process. Occasional jumps (occurring at a slower time scale) ride on top of a smooth drift process.

Gene regulation: In the gene regulation network (2.11), the active genes G are low in copy numbers and follow significantly discrete changes. However, the mRNA M and proteins P are relatively large in copy numbers and can be assumed to follow a diffusion process. A detailed account of such a hybrid model of a gene regulatory network can be found in [72] (switch plus diffusion) and [115, 175] (switch plus ODE).

Motivated by the above discussion, we will build in this chapter a general probabilistic framework to represent a range of processes in systems biology. This will allow us to review the formal relationships between different stochastic models referred to in the systems biology literature. As part of this review, we present a novel derivation of the differential Chapman–Kolmogorov equation for a general multidimensional Markov process made up of both continuous and jump processes. We start with the definition of a time derivative for a probability density but place no restrictions on the probability distribution; in particular, we do not assume it to be confined to a region that has a surface (on which the probability is zero). In our derivation, the master equation gives the jump part of the Markov process, while the Fokker–Planck equation gives the continuous part. We thereby sketch a "family tree" for stochastic models in systems biology, providing explicit derivations of their formal relationship and clarifying assumptions involved.

Key references in the area of stochastic modeling are the books [75], [54], [18], and [47]. Most stochastic models presented in these texts are derived on the basis of the Chapman–Kolmogorov equation (CKE), a consistency condition on Markov processes, in the form of a system of differential equations for the probability distribution. The system of differential equations takes the form of master equations for a jump Markov process and Fokker–Planck equations (FPE) for a continuous Markov process. For a detailed account of how this happens, read [75] and [54, 55]. For a Markov process that is made up of both jump and continuous parts, the differential equation takes the form of the differential Chapman–Kolmogorov equation (dCKE), which has been derived in [47]. The derivation is involved and requires the introduction of an arbitrary function, which leads to boundary restrictions on the probability distribution. As part of this review, we present a novel and more concise derivation of the dCKE. Since most of the mathematical foundations for stochastic models have been developed by physicists and mathematicians, we hope that our derivation makes the theory more accessible to the uninitiated researcher in the field of systems biology. We choose Markov processes as a framework, since more realistic approaches for modeling intracellular processes must take into account factors such as heterogeneity of the environment, macromolecular crowding [38, 138], and anomalous diffusion [60, 135, 136], to name a few. Anomalous diffusion is described by fractional Fokker–Planck equations [98]. Such treatments require advanced mathematical formalisms that are beyond the level assumed here.

The focus of the present chapter is to review the formal relationships between the equations referred to in the systems biology literature. We provide explicit derivations of their formal relationship and clarify assumptions involved in a common framework (Figure 8.3). We will end up in a family tree of stochastic models wherein the chemical master equation occupies a place for special jump processes. Similarly, the Fokker–Planck and the Langevin equations hold a place for diffusion processes. Such generalization provides a clearer picture of how the various stochastic approaches used in systems biology are related within a common framework.

8.1 Markov Processes

Markov processes form the basis for the vast majority of stochastic models of dynamical systems. The three books [47], [75], and [54] have become standard references for the application of Markov processes to biological and biochemical systems. At the center of a stochastic analysis is the Chapman–Kolmogorov equation (CKE), which describes the evolution of a Markov process over time. From the CKE stem three equations of practical importance: the master equation for jump Markov processes, the Fokker–Planck equation for continuous Markov processes, and the differential Chapman–Kolmogorov equation (dCKE) for processes made up both the continuous and jump parts. A nice mathematical (but nonbiological) account of these equations can also be found in [18]. Gardiner, van Kampen, and Gillespie take different approaches to derive these equations:

- Gillespie derives the FPE and the master equation independently from the CKE and for the one-dimensional case only in [54]. In [53, 55] he extends the derivations to multidimensional cases.

- Kampen [75] derives the master equation from the CKE for a one-dimensional Markov process. The FPE is given as an approximation of the master equation by approximating a jump process with a continuous one. The same approach is adopted in [18]. However, this should not mislead the reader to conclude that the FPE arises in this way. In fact, FPE is defined for a continuous Markov process.

- Gardiner [47] derives first the dCKE from the CKE for a multidimensional Markov process whose probability distribution is assumed to be contained in a closed surface. The FPE and the master equation are given as special cases of the dCKE.

We start with a review of notation for multidimensional probability theory required to read our proof. This is followed by a brief derivation of the CKE and its graphical interpretation. From the CKE we derive the dCKE and interpret its terms to show how the FPE and the master equation appear as special cases of the dCKE. Finally, it turns out that the CME is just a special form of the master equation for jump processes governed by chemical reactions. Note that while the derivations of the master equation and the Fokker–Planck equation in [54, 55] are for the one-dimensional case only, we here present a general treatment and derive all our results for multidimensional systems.

The probability distribution for an s-dimensional stochastic process

$$Y(t) = (Y_1(t), \ldots, Y_s(t))$$

is written as

$$\Pr\left[\bigcap_{i=1}^{s} y_i \leq Y_i(t) < y_i + \mathrm{d}y_i\right] = p(y_1, \ldots, y_s, t)\mathrm{d}y_1 \cdots \mathrm{d}y_s .$$

To simplify the notation, we use a short form,

$$\Pr\left[y \leq Y(t) < y + \mathrm{d}y\right] = p(y, t)\mathrm{d}y .$$

More useful will be the conditional probability density, $p(x, t \,|\, x', t')$, defined such that

$$\Pr\left[y \leq Y(t) < y + \mathrm{d}y \,|\, Y(t') = y'\right] = p(y, t \,|\, y', t')\mathrm{d}y .$$

When $t \geq t'$, $p(y, t \,|\, y', t')\mathrm{d}y$ is the transition probability from state y' at time t' to state y at time t. Since it is much easier to work with densities $p(\cdot)$ rather than probabilities $\Pr[\cdot]$, we shall use densities $p(\cdot)$, but abuse the terminology by referring to them as "probabilities."

Essentially, a Markov process is a stochastic process with a short-term memory. Mathematically, this means that the conditional probability of a state is determined entirely by the knowledge of the most recent state. Specifically, for any three successive times $t_0 \leq t \leq t + \Delta t$, one has

$$p(y, t + \Delta t \,|\, y', t; y_0, t_0) = p(y, t + \Delta t \,|\, y', t),$$

where the conditional probability of y at $t + \Delta t$ is uniquely determined by the most recent state y' at t and is not affected by any knowledge of the initial state y_0 at t_0. This Markov property is assumed to hold for any number of successive time intervals. To see how powerful this property is, let us consider the factorization of the joint probability

$$p(y, t + \Delta t ; y', t) = p(y, t + \Delta t \,|\, y', t)p(y, t) .$$

Making both sides conditional on (y_0, t_0) will modify this equation to

$$p(y, t + \Delta t ; y', t \,|\, y_0, t_0) = p(y, t + \Delta t \,|\, y', t ; y_0, t_0)p(y', t \,|\, y_0, t_0),$$

which, by the Markov property, reduces to

$$p(y, t + \Delta t ; y', t \,|\, y_0, t_0) = p(y, t + \Delta t \,|\, y', t)p(y', t \,|\, y_0, t_0) . \qquad (8.1)$$

The last equation shows that the joint probability can be expressed in terms of transition probabilities. Recall the following rule for joint probabilities,

$$p(y) = \int dy' \, p(y, y'), \qquad (8.2)$$

which says that summing a joint probability over all values of one of the variables eliminates that variable. Now integrating (8.1) over y' and using (8.2), we arrive at the Chapman–Kolmogorov equation (CKE) [47]:

$$p(y, t + \Delta t \,|\, y_0, t_0) = \int dy' \, p(y, t + \Delta t \,|\, y', t)p(y', t \,|\, y_0, t_0) . \qquad (8.3)$$

This equation expresses the probability of a transition $(y_0 \to y)$ as the summation of probabilities of all transitions $(y_0 \to y' \to y)$ via intermediate states y'. Figure 5.2 illustrates the basic notion of a Markov process for which the CKE provides the stochastic formalism. When the initial condition (y_0, t_0) is fixed, which is assumed here, the transition probability conditioned on (y_0, t_0) is the same as the state probability:

$$p(y, t) = p(y, t \,|\, y_0, t_0) .$$

8.2 Derivation of the dCKE

The CKE serves as a description of a general Markov process, but cannot be used to determine the temporal evolution of the probability. Here we derive from the CKE a differential equation that will be more useful in terms of describing the dynamics of the stochastic process. Referred to as the "differential Chapman–Kolmogorov equation" (dCKE) by Gardiner [47], this equation contains the CME as a special case. This derivation is for a multidimensional Markov process. We start with the definition of a time derivative for a probability density but place no restrictions on the probability distribution. Gardiner [47] instead starts with the expectation of an arbitrary function, which results in integration by parts and consequently the need to assume that the probability density vanishes on the surface of a region to which the process is confined. We do not need such as assumption because of the simplicity of our approach. The master equation gives the jump part of the Markov process, while the Fokker–Planck equation gives the continuous part.

Consider the time derivative of the transition probability

$$\frac{\partial p(y,t)}{\partial t} = \lim_{\Delta t \to 0} \frac{1}{\Delta t} \left\{ p(y, t + \Delta t) - p(y, t) \right\}, \tag{8.4}$$

where differentiability of the transition probability with respect to time is assumed. Employing the CKE (8.3) and the normalization condition

$$\int dy' \, p(y', t + \Delta t \,|\, y, t) = 1 \,,$$

since $p(y', t + \Delta t \,|\, y, t)$ is a probability, (8.4) can be rewritten as

$$\frac{\partial p(y,t)}{\partial t} = \lim_{\Delta t \to 0} \frac{1}{\Delta t} \int dy' \left\{ p(y, t + \Delta t \,|\, y', t) p(y', t) - p(y', t + \Delta t \,|\, y, t) p(y, t) \right\}.$$

Let us divide the region of integration into two regions based on an arbitrarily small parameter $\epsilon > 0$. The first region $\|y - y'\| < \epsilon$ contributes a continuous-state process, whereas the second region $\|y - y'\| \geq \epsilon$ contributes to a jump process. Here $\|\cdot\|$ denotes a suitable vector norm. Let I_R denote the right side of the above equation in region R of the state space, that is,

$$I_R = \lim_{\Delta t \to 0} \frac{1}{\Delta t} \int_R dy' \left\{ p(y, t + \Delta t \,|\, y', t) p(y', t) - p(y', t + \Delta t \,|\, y, t) p(y, t) \right\}, \tag{8.5}$$

then the derivative (8.6) in the whole region of the state space can be expressed by

$$\frac{\partial p(y,t)}{\partial t} = I_{\|y-y'\|<\epsilon} + I_{\|y-y'\|\geq\epsilon}, \tag{8.6}$$

The integrand of $I_{\|y-y'\|<\epsilon}$ can be expanded in powers of $y - y'$ using a Taylor expansion. Setting $y - y' = q$, we can write

$$I_{\|q\|<\epsilon} = \lim_{\Delta t \to 0} \frac{1}{\Delta t} \int_{\|q\|<\epsilon} dq \left\{ p(y, t + \Delta t \,|\, y - q, t) p(y - q, t) \right.$$

$$\left. - p(y - q, t + \Delta t \,|\, y, t) p(y, t) \right\}. \qquad (8.7)$$

In order to expand the integrand more easily into a Taylor series, let us define a function

$$f(y; q) \triangleq p(y + q, t + \Delta t \,|\, y, t) p(y, t),$$

so that the integrand in (8.7) becomes $f(y - q; q) - f(y; -q)$, which, after a Taylor expansion, becomes

$$-f(y; -q) + f(y; q) - \sum_i q_i \frac{\partial f(y; q)}{\partial y_i} + \frac{1}{2} \sum_{i,j} q_i q_j \frac{\partial^2 f(y; q)}{\partial y_i \partial y_j} + o\left(\|qq^T\|\right).$$

The integrals of the first two terms cancel because of the symmetry

$$\int f(y; q) dq = \int f(y; -q) dq$$

when the integral is over all the positive and negative values of q in the region. Thus, we have

$$I_{\|q\|<\epsilon} = \lim_{\Delta t \to 0} \frac{1}{\Delta t} \int_{\|q\|<\epsilon} dq \left\{ o\left(\|qq^T\|\right) \right.$$

$$- \sum_i q_i \frac{\partial \left[p(y + q, t + \Delta t \,|\, y, t) p(y, t) \right]}{\partial y_i}$$

$$\left. + \frac{1}{2} \sum_{i,j} q_i q_j \frac{\partial^2 \left[p(y + q, t + \Delta t \,|\, y, t) p(y, t) \right]}{\partial y_i \partial y_j} \right\}.$$

For the state increments $\Delta Y_i = Y_i(t + \Delta t) - Y_i(t)$, recognizing the (conditional) expectations

$$\langle \Delta Y_i \rangle_y = \int_{\|q\|<\epsilon} dq \, q_i \, p(y + q, t + \Delta t \,|\, y, t)$$

and

$$\langle \Delta Y_i \Delta Y_j \rangle_y = \int_{\|q\|<\epsilon} dq \, q_i q_j \, p(y+q, t+\Delta t \,|\, y, t),$$

we refer to the differentiability conditions for continuous processes, i.e., $\|y - y'\| < \epsilon$ [47, Section 3.4]:

$$\lim_{\Delta t \to 0} \frac{\langle \Delta Y_i \rangle_y}{\Delta t} = A_i(y, t) + o(\epsilon), \tag{8.8}$$

$$\lim_{\Delta t \to 0} \frac{\langle \Delta Y_i \Delta Y_j \rangle_y}{\Delta t} = B_{ij}(y, t) + o(\epsilon). \tag{8.9}$$

The higher-order terms involve higher-order coefficients, which must vanish. To see this, for the third-order coefficient, we have

$$\lim_{\Delta t \to 0} \frac{1}{\Delta t} \int_{\|q\|<\epsilon} dq \, q_i q_j q_k \, p(y+q, t+\Delta t \,|\, y, t) = C_{ijk}(y, t) + o(\epsilon).$$

However,

$$\lim_{\Delta t \to 0} \frac{1}{\Delta t} \int_{\|q\|<\epsilon} dq \, q_i q_j q_k \, p(y+q, t+\Delta t \,|\, y, t)$$

$$\leq \|q\| \lim_{\Delta t \to 0} \frac{1}{\Delta t} \int_{\|q\|<\epsilon} dq \, q_i q_j \, p(y+q, t+\Delta t \,|\, y, t)$$

$$\leq \epsilon \left[B_{ij}(y, t) + o(\epsilon) \right] \leq o(\epsilon).$$

Hence $C(y, t)$ must vanish. The vanishing of higher-order coefficients follows immediately. The coefficient $A(y, t)$ turns out to be the drift rate, and $B(y, t)$ the diffusion rate. Comparison of (8.8) with (5.28) suggests that $A(y, t)$ is analogous to the drift rate $A(n)$ defined in (5.29). Similarly, comparison of (8.9) with (5.28) suggests that the matrix $B(y, t)$ is analogous to the diffusion rate $B(n)$ defined in (5.31). The above analogy can be informally explained in the following manner. Given $Y(t) = y$, the state increment vector $Y(t + dt) - Y(t)$ for a continuous process has a mean approaching $A(y, t)dt$ and a covariance approaching $B(y, t)dt$, as ϵ approaches zero. This suggests the following update rule, for $\epsilon \to 0$ and under assumptions given in [47, Section 3.5.2]:

$$Y(t + dt) - Y(t) = A(Y(t), t)dt + [B(Y(t), t)]^{1/2} \, d\mathcal{W} \tag{8.10}$$

which has the same (alternative) form as the Langevin equation in (5.39).

This also explains why we have chosen the same notation for these coefficients appearing in different contexts. We remark here that (8.8) and (8.9) are postulated here for mathematical convenience. A more rigorous justification is given in [55]. Subject to the differentiability conditions (8.8) and (8.9), we see that as $\epsilon \to 0$,

$$I_{\|y-y'\|<\epsilon} \to -\frac{\partial}{\partial y}\left[A(y,t)p(y,t)\right] + \frac{1}{2}\frac{\partial^2}{\partial y \partial y^T}\left[B(y,t)p(y,t)\right]. \qquad (8.11)$$

Next we work out the jump probability rate $I_{\|y-y'\|\geq\epsilon}$ defined by (8.5). We will use the differentiability condition for jump processes, i.e., $\|y - y'\| \geq \epsilon$ [47, Section 3.4]:

$$\lim_{\Delta t \to 0} \frac{1}{\Delta t} p(y, t + \Delta t \,|\, y', t) = W(y \,|\, y', t),$$

where $W(y \,|\, y', t)$ is called the transition rate for the jump $(y' \to y)$. Subject to this condition, we see that as $\epsilon \to 0$, the region of integration approaches the full state space, leading to

$$I_{\|y-y'\|\geq\epsilon} \to \int dy' \left[W(y \,|\, y', t)p(y', t) - W(y' \,|\, y, t)p(y, t)\right]. \qquad (8.12)$$

Adding (8.11) and (8.12), we can rewrite (8.6) to arrive at the dCKE:

$$\frac{\partial}{\partial t}p(y,t) = -\frac{\partial}{\partial y}\left[A(y,t)p(y,t)\right] + \frac{1}{2}\frac{\partial^2}{\partial y \partial y^T}\left[B(y,t)p(y,t)\right]$$
$$+ \int dy' \left[W(y \,|\, y', t)p(y', t) - W(y' \,|\, y, t)p(y, t)\right]. \qquad (8.13)$$

We now have a differential equation characterizing the dynamics of the probability distribution $p(y, t)$, that is, the probability of a state at any time, starting from a given initial probability distribution. This completes our derivation of the differential Chapman–Kolmogorov equation. The following section will classify Markov processes based on this dCKE. This is followed by a derivation of the chemical master equation and its use in systems biology.

8.3 Classification of Markov Processes

Being a linear differential equation, the dCKE is more convenient for mathematical treatment than the original CKE. More importantly, it has a more direct physical interpretation. The coefficients $A(y, t)$, $B(y, t)$, and $W(y' \,|\, y, t)$ are specified by the system under consideration, and thus the solution of the dCKE gives the probability distribution for the state of the given system [75]. The original CKE, on the other hand, has no specific information about any particular Markov process. We now interpret the different terms of (8.13). Following [18, 47], we first consider the case

$$B_{ij}(y,t) = W(y\,|\,y',t) = W(y'\,|\,y,t) = 0,$$

reducing the dCKE to

$$\frac{\partial}{\partial t}p(y,t) = -\frac{\partial}{\partial y}\Big[A(y,t)p(y,t)\Big],$$

which is a special case of the *Liouville equation* describing a deterministic motion (see [47, Section 3.5.3]):

$$\frac{\mathrm{d}}{\mathrm{d}t}y(t) = A(y,t).$$

Next, if $A(y,t) = B(y,t) = 0$, the CKE reduces to

$$\frac{\partial}{\partial t}p(y,t) = \int \mathrm{d}y'\,\Big[W(y\,|\,y',t)p(y',t) - W(y\,|\,y',t)p(xy,t)\Big]. \qquad (8.14)$$

This is the master equation describing a jump Markov process with discontinuous (or discrete) sample paths. An example of a jump process is the time-dependent mRNA abundance in a gene regulatory network.

Next, if $W(y\,|\,y',t) = W(y'\,|\,y,t) = 0$, the CKE reduces to

$$\frac{\partial}{\partial t}p(y,t) = -\frac{\partial}{\partial y}\Big[A(y,t)p(y,t)\Big] + \frac{1}{2}\frac{\partial^2}{\partial y\partial y^T}\Big[B(y,t)p(y,t)\Big],$$

which is the Fokker–Planck equation (FPE) and is equivalent to the Langevin equation (8.10) under the conditions given in [47, 55, 75]. The corresponding process is known as a "diffusion process" (a Markov process with continuous sample paths). An example of a diffusion process is the metabolite concentration of a metabolic network. This shows that the FPE is originally defined for a continuous process. However, the FPE can also arise as an approximation of the master equation when the jumps of the corresponding discrete process are assumed to be small [75, 79].

Finally, we consider the case without diffusion, that is, $B(y,t) = 0$, which leads us to

$$\frac{\partial}{\partial t}p(y,t) = -\frac{\partial}{\partial y}\Big[A(y,t)p(y,t)\Big] + \int \mathrm{d}y'\,\Big[W(y\,|\,y',t)p(y',t) - W(y'\,|\,y,t)p(y,t)\Big].$$

which is called the "Liouville master equation" (LME) in [18, Chapter 1] and describes a piecewise deterministic process with sample paths consisting of smooth deterministic pieces interrupted by instantaneous jumps. One way in which the LME arises is through the approximation of an originally jump Markov process by a hybrid process with discrete and continuous parts [93, 115].

In the most general case, in which none of the quantities $A(y,t)$, $B(y,t)$, and $W(y'\,|\,y,t)$ vanish, the dCKE may describe a process whose sample paths are piecewise continuous, made up of pieces that correspond to a diffusion process with a nonzero drift, onto which is superimposed a fluctuating part.

8.4 Chemical Master Equation

Consider a discrete-state Markov process. The master equation for this process can be obtained from (8.14), to give

$$\frac{\partial P(n,t)}{\partial t} = \sum_{n'}\Big[W(n\,|\,n',t)P(n',t) - W(n'\,|\,n,t)P(n,t)\Big],$$

where n' is the intermediate state, and n the final state. Since $P(n,t)$ is a probability (and not a density), the integral $\int$ in (8.14) has been replaced by the summation $\sum_{n'}$. We can rewrite this equation in terms of jumps $q = n - n'$,

$$\frac{\partial P(n,t)}{\partial t} = \sum_{q}\Big[W(n\,|\,n-q,t)P(n-q,t) - W(n+q\,|\,n,t)P(n,t)\Big], \quad (8.15)$$

where we have used the symmetry $\sum_q \phi(-q) = \sum_q \phi(q)$, for an arbitrary function $\phi(\cdot)$, when writing the second summand. Now consider an s-component and r-reaction biochemical system. Let i label the different components (chemical species) and j label different reaction channels. The copy number of the ith component at the variable time t will be denoted by $Y_i(t)$, which takes values n_i from the set of whole numbers. Each occurrence of the jth reaction channel changes the copy number n_i of the ith component by an amount S_{ij}, an element of the stoichiometry matrix S. It is assumed that the species are distributed homogeneously (well mixed) in a closed system of constant volume Ω at a constant temperature. This essentially assumes that changes depend only on the current state (Markov property) and that we can avoid spatial considerations [36, 47, 75] and macromolecular crowding [63]. However, since diffusion may not always be rapid, spatial considerations become important in dealing with intracellular processes [35, 80, 85]. Here we are interested in a stochastic formulation that dates back to the initial work by Kramers [84]. Under the stated assumptions, the vector

$$Y(t) = (Y_1(t), \ldots, Y_s(t))$$

taking values $n = (n_1, \ldots, n_s)$ is a continuous-time Markov process. The jump sizes are determined by the stoichiometry and molecularity of the reactions and therefore can take values only from the set $\{S._1, \ldots, S._r\}$ of the elementary changes. Thus, for our system of chemical reactions, (8.15) becomes

$$\frac{\partial P(n, t)}{\partial t} = \sum_{j=1}^{r} \Big[W(n \mid n - S._j, t) P(n - S._j, t) - W(n + S._j \mid n, t) P(n, t) \Big].$$

Since $S._j$ is uniquely defined for a reaction R_j, we can recognize the associated transition rate as a reaction propensity (defined in Chapter 5)

$$a_j(n) = W(n + S._j \mid n, t)$$

which means that the above master equation is simply the chemical master equation:

$$\frac{\partial}{\partial t} P(n, t) = \sum_{j=1}^{r} [a_j(n - S._j) P(n - S._j, t) - a_j(n) P(n, t)].$$

This shows that the CME is just a special form of the master equation for jump processes governed by chemical reactions.

8.5 Stochastic Family Tree

The formal relationships between the equations referred to in the systems biology literature, and reviewed in this chapter, can be depicted as a "family tree" shown in Figure 8.3. This figure illustrates the links between different stochastic models, between different simulation methods, and between modeling and simulation.

Given that models of cellular systems are often modest, compared to what nature presents us with, it is helpful to discuss the role of (mathematical) modeling. Mathematical modeling serves several purposes that go beyond the usual association of models with prediction. In an essay, Arthur D. Lander [87] referred to an analysis of Epstein [41], which lists numerous reasons for modeling, from which we consider the following most important for systems biology:

- Explain (very distinct from predict)

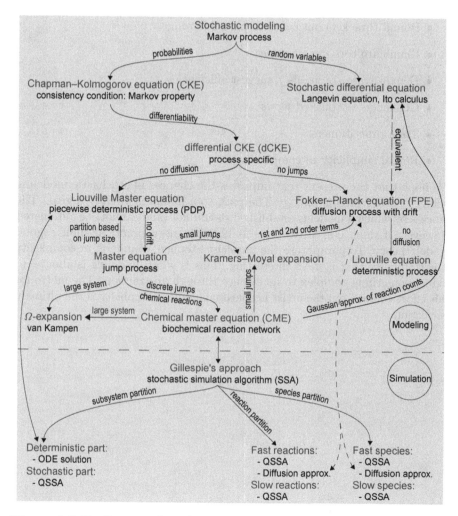

Figure 8.3 Family tree of stochastic processes: Interrelationships for various stochastic approaches. QSSA stands for Quasi-Steady-State Assumption. The figure first appeared in our earlier work [158].

- Guide data collection

- Illuminate core dynamics

- Suggest dynamical analogies

- Discover new questions

- Promote a scientific habit of mind

- Bound (bracket) outcomes to plausible ranges

- Illuminate core uncertainties

- Demonstrate trade-offs / suggest efficiencies

- Challenge established views

- Train practitioners

- Reveal simplicity in complexity

We hope that the present text improves the chances of stochastic modeling being used in systems biology The path to a model is already a goal. The process by which a model is established should not (and cannot) be automated. However, if done correctly, there is nothing more practical than a good model/theory. To paraphrase the physicist Stephen Weinberg, in complexity it is simplicity that is most interesting. This is what modeling is about—to reduce something complex to its essence through abstraction. The motto for this effort comes from the artist Leonardo da Vinci: Simplicity is the ultimate perfection.

Chapter 9

Wet-Lab Experiments and Noise

In this chapter we review selected publications on noise and stochastic modeling that are linked to experimental studies. Due to the wide range of experimental technologies used to generate data, and because the importance of this to the analysis, we cannot reproduce these studies in a book like this. The selection of a few papers is to demonstrate the relevance of noise and stochastic modeling to state-of-the-art molecular and cell biology.

9.1 Reviews

In a recent *Nature* review [32], Avigdor Eldar and Michael B. Elowitz highlight the functional roles of noise in genetic circuits. They demonstrate that stochastic fluctuations, or "noise," in the levels of components in genetic circuits is not just a nuisance, but has in fact a purposeful role in the regulation of cellular functions. More specifically, their review focuses on the fact that noise (i) enables physiological regulation mechanism, (ii) permits probabilistic differentiation strategies in microbial to multicellular organisms at the population level, and (iii) facilitates evolutionary adaptation and developmental evolution. In addition to a concise summary of the literature and key experimental works, the review is an excellent demonstration of how theoretical (here stochastic) approaches can be combined with state-of-the-art technologies to generate data. The small size of prokaryotic cells and very low copy numbers of molecules are convincing reminders of what it takes to quantify subcellular processes in individual cells. Despite the impressive array of convincing examples, Eldar and Elowitz conclude with a summary of remaining challenges and open questions, including the question of how noise emerges. They point toward microRNAs as a possible important player, adding to the considerable attention microRNAs have received very recently. Their review covers the subcellular level up to the level of physiological processes such as metabolic networks and hormone regulation of tissues. At this level, a question remains as to which systems use noise in physiological processes, something that would require technologies to monitor a diverse range of biochemical reactions in individual living cells. A third challenge, again requiring further advances in single-cell technologies, is probabilistic differentiation in stem cells to understand both how and why cells switch dynamically among states or substates.

Michael Elowitz is a leading figure in the study of noise in gene expression. Together with his coworkers, a key contribution he has made is the discovery that noise can have a purposeful role in cells. In [31] it is shown that correlations in gene expression noise could provide a noninvasive means to probe the activity states of regulatory links. The underlying idea is that gene regulatory interactions are context-dependent, active in some cellular states but not in others, and that gene-expression noise propagates only through active regulatory links. The authors show that single-cell time-lapse microscopy can be used to discriminate between active regulatory connections and extrinsic noise. The experimental work, focusing on galactose metabolism genes in *Escherichia coli*, is complemented by a mathematical analysis and a model of three differential equations and extrinsic noise signals modeled as an Ornstein–Uhlenbeck process.

Noise can be used to analyze the activity of gene regulatory interactions as depicted in Figure 9.1. "Extrinsic noise" is here defined as the overall rate of expression of all genes, such as fluctuating numbers of ribosomes, polymerase, and variations in cell size, affecting the expression of many genes. This global, extrinsic noise is thus to be distinguished from "intrinsic noise" in the expression of individual genes. Because it takes time for protein concentrations to build up, gene regulation occurs with a delay. The sign of the delay between fluctuation in regulator concentration and its effect on target protein levels thus provides information about the causal direction of the link. Since such delays do not occur for global extrinsic noise, which affects all genes simultaneously, Elowitz and colleagues developed a strategy to decouple extrinsic noise correlations from regulatory correlations.

In [141], "extrinsic noise" is defined as nonspecific fluctuations affecting many system components and having long duration compared to the cell cycle. Extrinsic fluctuations are then referred to as colored. The notion of white or colored noise has its origin in the engineering sciences and is linked to the spectral density (power distribution in the frequency spectrum) of a noise signal. White noise is a process, named in analogy to white light, with a flat frequency spectrum (white light being a uniform mix of wavelengths). In contrast to white noise, colored noise does not vary completely randomly. A random process whose power spectral density is not white or nearly white is known in the engineering sciences as "colored noise".

The trick to distinguish between intrinsic and extrinsic noise in experiments is to create a copy of the network in the same cellular environment as the original network. The idea, going back to [39], is that intrinsic variables that define the copy numbers of components of the network are specific to that network and thus differ between the copies of the network, while extrinsic

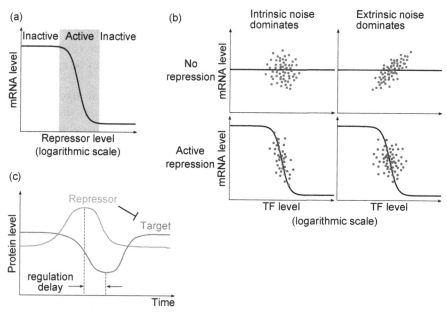

Figure 9.1 Noise-based activity analysis of gene regulatory interactions. (a) Target gene expression versus repressor concentration. In the *active* region (in the middle), changes in repressor concentration cause changes in target gene expression. The link is *inactive'* for higher (right) and lower (left) repressor levels. (b) Noise can produce different types of static correlations between transcription factor concentration and target gene expression. In each plot, dots represent individual cells. Top plots show correlations without an active regulatory link, whereas bottom plots show correlations with active repression. The mean gene regulation function is shown as a solid line. Two noise regimes are considered in which either intrinsic (uncorrelated) or extrinsic (correlated) noise dominates. Active repression causes anticorrelation between the transcription factor and its target. Intrinsic noise decorrelates the two, and extrinsic noise causes positive correlations even without active regulation. Thus, correlations derived from static snapshots are ambiguous because high levels of extrinsic noise can conceal the anticorrelation expected from repression. (c) Temporal gene expression patterns for a repressor and its target showing anticorrelations at a delay time. Figure modified from [31, Figure 1].

variables describe molecules that affect equally each copy of the network. For example, the abundance of transcribing RNA polymerase is an intrinsic variable, which differs for each copy of the network. On the other hand, the abundance of cytosolic RNA polymerase is an extrinsic variable for which all copies of the network are exposed to the same numbers. Measured fluctuations of the intrinsic variable will have intrinsic and extrinsic components: intrinsic variables are themselves part of a stochastic system, and that system interacts with other stochastic systems. The term "noise" is used in [141] to mean exclusively an empirical measure of stochasticity, usually the coefficient of variation. Experimentally, the relative number of proteins can be quantified in living cells using fluorescent proteins.

The authors extend the standard stochastic simulation algorithm to include extrinsic fluctuations and show that these fluctuations affect mean protein numbers. Because extrinsic fluctuations typically cause fluctuations in the parameters of a network, their extension of the Gillespie algorithm includes discontinuous, time-varying parameters. The study considers a simple model of gene expression that includes promoter activation, transcription, translation, and degradation. The rate of translation is a function of the number of free ribosomes and is thus an extrinsic variable, fluctuating with an average lifetime that is nonzero (thus "colored").

Biological noise plays an import role in cell fate decisions generating nongenetic cellular diversity, which may be critical for development, resource utilization and survival in a fluctuating environment. In a recent review [10], the role of regulatory network structure and molecular noise is highlighted using several examples of stochastic cellular decision making from viruses, bacteria, yeast, lower metazoans, and mammals. The authors consider cellular decision as one of the three key processes underlying development at various scales of biological organization. Environmental sensing and cell-cell communication are the other two key processes. The networks behind cellular decision making are characterized by (1) feedback loops for stability of cellular decisions (positive feedback) and reversibility (negative feedback), (2) bistability or excitable dynamics, and (3) intrinsic molecular noise inducing transitions between steady states in bistable systems and transient excursions of gene expression in excitable systems. An interesting proposition by the authors is that efforts to reduce intrinsic noise in experimental setups will always be in vain. Intrinsic noise is unavoidable by its very nature. Thus noise should be better understood, incorporated in modeling and subsequently exploited to control cell fate decisions.

Other important processes wherin noise plays a role include development and pattern formation [88].

9.2 Further Reading

Here is a list of articles and books for further reading.

Modeling Approaches

- A. Elston: Physics to pharmacology. *Rep. Prog. Phys.*, 74(2011), e016601.

- D. Pe'er and N. Hacohen: Principles and strategies for developing network models in cancer. *Cell*, 144(2011), 864–873.

- W. W. Chen, M. Niepel and P. K. Sorger: Classic and contemporary approaches to modeling biochemical reactions. *Genes. Dev.*, 1(2010):1861–1875.

- P. K. Kreeger and D. A. Lauffenburger: Cancer systems biology: a network modeling perspective. *Carcinogenesis*, 31(2010), 2–8.

- G. Karlebach and R. Shamir: Modeling and analysis of gene regulatory networks. *Nature Reviews—Molecular Cell Biology*, 9(2008), 770–780.

- E.J. Crampin, S. Schnell, P.E. McSharry: Mathematical and computational techniques to deduce complex biochemical reaction mechanisms. *Progress in Biophysics & Molecular Biology*, 86(2004), 77–112.

- J.J. Tyson, K.C. Chen, B. Novak: Sniffers, buzzers, toggles and blinkers: dynamics of regulatory and signaling pathways in the cell. *Current Opinion in Cell Biology*, 15(2003), 221–231.

Stochastic Modeling

- R. Cheong, S. Paliwal, and A. Levchenko: Models at the single cell level. *WIREs. Syst. Biol. Med.*, 2(2010), 34–48.

- S. J. Altschuler and L. F. Wu: Cellular heterogeneity: Do differences make a difference? *Cell*, 141(2010), 559–563.

- R. J. Johnston and C. Desplan: Stochastic mechanisms of cell fate specification that yield random or robust outcomes. *Annu. Rev. Cell. Dev. Biol.*, 26(2010), 689–719.

- D. J. Wilkinson: Stochastic modelling for quantitative description of heterogeneous biological systems. *Nat. Rev. Genet.*, 10(2009), 122–133.

- J. Pahle: Biochemical simulations: stochastic, approximate stochastic and hybrid approaches. *Brief. Bioinform.*, 10(2009), 53–64.

- D. J. Higham: Modeling and simulating chemical reactions. *SIAM Review*, 50(2008), 347–368.

- D. T. Gillespie: Stochastic simulation of chemical kinetics. *Annu. Rev. Phys. Chem.*, 58(2007), 35–55.

- N. Maheshri and E. K. O'Shea: Living with noisy genes: How cells function reliably with inherent variability in gene expression. *Annu. Rev. Biophys. Biomol. Struct.*, 36(2007), 413–434.

- C. Blomberg: Fluctuations for good and bad: The role of noise in living systems. *Physics of Life Reviews*, 3(2006), 133–161.

- M. Kaern, T. C. Elston, W. J. Blake and J. J. Collins: Stochasticity in gene expression: from theories to phenotypes. *Nat. Rev. Genet.*, 6(2005), 451–464.

- J. M. Raser and E. K. O'Shea: Noise in gene expression: Origins, consequences, and control. *Science*, 309(2005), 2010–2013.

- J. Paulsson: Models of stochastic gene expression. *Phys. Life. Rev.*, 2(2005), 157–75.

- T. Turner, S. Schnell and K. Burrage: Stochastic approaches for modelling in vivo reactions. *Comput. Biol. Chem.*, 28(2004), 165–178.

- K. S. Dorman, J. S. Sinsheimer and K. Lange: In the garden of branching processes. *SIAM Review*, 46(2004), 202–229.

Cellular Oscillations

- J, J. Ferrell, T.-C. Tsai and Q. Yang: Modeling the cell cycle: Why do certain circuits oscillate? *Cell*, 144(2011), 874–885.

- B. Novák and J.J. Tyson: Design principles of biochemical oscillators. *Nature Reviews—Molecular Cell Biology*, 9(2008), 981–991.

- T. Roenneberg, E. Mendoza: Modeling biological rhythms. *Current Biology*, 18(2008), 826–835.

- F. Levi, U. Schibler: Circadian rhythms: Mechanisms and therapeutic implications. *Annu. Rev. Pharmacol. Toxicol.*, 47(2007), 593–628.

- A. Goldbeter: Computational approaches to cellular rhythms. *Nature*, 420(2002), 238–245.

Cell Signaling

- J. Muñoz-García and B. N. Kholodenko: Signalling over a distance: gradient patterns and phosphorylation waves within single cells. *Biochem. Soc. Trans.*, 38(2010), 1235–1241.

- B. N. Kholodenko, J. F. Hancock and W. Kolch: Signalling ballet in space and time. *Nat. Rev. Mol. Cell. Biol.*, 11(2010), 414–26.

- O. Brandman and T. Meyer: Feedback loops shape cellular signals in space and time. *Science*, 322(2008), 390–395.

- B.N. Kholodenko: Cell-signaling dynamics in time and space. *Molecular Cell Biology*, 7(2006), 165–176.

- K.A. Janes, M.B. Yaffe: Data-driven modeling of signal-transduction networks. *Nature Cell Biology*, 8(2006), 820–828.

- B. B. Aldridge, J.M. Burke, D.A. Lauffenburger, P. Sorger: Physicochemical modeling of cell signaling pathways. *Nature Cell Biology*, 8(2006), 1195–1203.

- R. Heinrich, B.G. Neel, T.A. Rapoport: Mathematical models of protein kinase signal transduction. *Molecular Cell*, 9(2002), 957–970.

- D. A. Lauffenburger and J. J. Linderman: *Receptors: Models for Binding, Trafficking, and Signalling*. Oxford University Press, (1993); 2nd printing (1996).

Diagrammatic Modeling

- N.Le Novere *et al.*: The systems biology graphical notation. *Nature Biotechnology*, 27(2009), 735–741.

- K.W. Kohn, M.I. Aladjem, J.N. Weinstein, Y. Pommier: Molecular interaction maps of bioregulatory networks: A general rubric for systems biology. *Molecular Biology of the Cell*, 17(2006), 1–13.

- H. Kitano, A. Funahashi, Y. Matsuoka, K. Oda: Using process diagramms for the graphical representation of biological networks. *Nature Biotechnology*, 23(2005), 961–966.

Handling Experimental Data

- S. Bandara, J.P. Schlöder, R. Eils, H.G. Bock, T. Meyer: optimal experimental design for parameter estimation of a cell signaling model. *PLoS. Comput. Biol.*, 5(2009), e1000558.

- A. Raue, C. Kreutz, T. Maiwald, J. Bachmann, M. Schilling, U. Klingmüller, J. Timmer: Structural and practical identifiability analysis of partially observed dynamical models by exploiting theprofile likelihood. *Bioinformatics*, 25(2009), 1923–1929.

- T. Maiwald, J. Timmer: Dynamical modeling and multi-experiment fitting with PottersWheel. *Bioinformatics*, 24(2008), 2037–2043.

- E. Balsa-Canto, A.A. Alonso, J.R. Banga: Computational procedures for optimal experimental design in biological systems. *IET Systems Biology*, 2(2008), 163–172.

- S.J. Wilkinson, N. Benson and D.B. Kell: Proximate parameter tuning for biochemical networks with uncertain. *Mol. BioSyst.*, 4(2008), 74–97.

- R.N. Gutenkunst, J.J. Waterfall, F.P. Casey, K.S. Brown, C.R. Myers, J.P. Sethna: Universally sloppy parameter sensitivities in systems biology models. *PLoS. Comput. Biol.*, 3(2007), e189.

- S. Hengl, C. Kreutz, J. Timmer, T. Maiwald: Data-based identifiability analysis of non-linear dynamical models. *Bioinformatics*, 23(2007), 2612–2618.

- M. Rodriguez-Fernandez, J.A. Egea, J.R. Banga: Novel metaheuristic for parameter estimation in nonlinear dynamic biological systems. *BMC Bioinformatics*, 7(2006), e483.

- M. Schilling, T. Maiwald, S. Bohl, M. Kollmann, C. Kreutz, J. Timmer, U. Klingmüller: Quantitative data generation for systems biology: the impact of randomisation, calibrators and normalizers. *Syst. Biol. (Stevenage)*, 152(2005), 193–200.

Personal Perspectives

- J.E. Ferrell Jr: Q&A: Systems biology. *Journal of Biology*, 8:2(2009).

- O. Wolkenhauer, M. Mesarovic: Feedback dynamics and cell function: Why systems biology is called Systems Biology. *Mol. BioSyst.*, 1(2005), 14–16.

- L. Hood, J.R. Heath, M.E. Phelps, B. Lin: Systems biology and new technologies enable predictive and preventative medicine. *Science*, 306(2004), 640–643.

- H.S. Wiley: Systems biology: Beyond the buzz. *The Scientist*, 20(2006), e52.

- H. Kitano: Computational systems biology. *Nature*, 420(2002), 206–210.

- Y. Lazebnik: Can a biologist fix a radio? Or, what I learned while studying apoptosis. *Cancer Cell*, 2(2002), 179–182.

References

[1] IUPAC Compendium of Chemical Terminology—the Gold Book.

[2] Virtual Laboratories in Probability and Statistics. URL http://www.math.uah.edu/stat/.

[3] T.-H. Ahn, L. T. Watson, Y. Cao, C. A. Shaffer, and W. T. Baumann. Cell cycle modeling for budding yeast with stochastic simulation algorithms. *CMES*, 51(1):27–52, 2009. doi: 10.3970/cmes.2009.051.027.

[4] B. Alberts, A. Johnson, J. Lewis, M. Raff, K. Roberts, and P. Walker. *Molecular Biology of the Cell.* Garland Publishers, New York, 4th edition, 2002.

[5] L. J. Allen. *An Introduction to Stochastic Processes with Applications to Biology.* Prentice Hall, Apr. 2003. ISBN 0130352187.

[6] J. Ansel, H. Bottin, C. Rodriguez-Beltran, C. Damon, M. Nagarajan, S. Fehrmann, J. François, and G. Yvert. Cell-to-cell stochastic variation in gene expression is a complex genetic trait. *PLoS Genet.*, 4(4):e1000049, Apr 2008. doi: 10.1371/journal.pgen.1000049.

[7] A. Arkin, J. Ross, and H. H. McAdams. Stochastic kinetic analysis of developmental pathway bifurcation in phage λ-infected *Escherichia coli* cells. *Genetics*, 149(4):1633–1648, 1998. URL http://www.genetics.org/cgi/content/abstract/149/4/1633.

[8] M. N. Artyomov, J. Das, M. Kardar, and A. K. Chakraborty. Purely stochastic binary decisions in cell signaling models without underlying deterministic bistabilities. *Proc. Natl. Acad. Sci. U.S.A.*, 104(48):18958–18963, Nov. 2007. doi: 10.1073/pnas.0706110104.

[9] S. Aumaître, K. Mallick, and F. Pétrélis. Noise-induced bifurcations, multiscaling and on–off intermittency. *Journal of Statistical Mechanics: Theory and Experiment*, 2007(07):P07016, 2007. doi: 10.1088/1742-5468/2007/07/P07016.

[10] G. Balázsi, A. van Oudenaarden, and J. Collins. Cellular decision making and biological noise: From microbes to mammals. *Cell*, 144(6): 910–925, Mar. 2011. ISSN 0092-8674. doi: 10.1016/j.cell.2011.01.030.

[11] D. A. Beard and H. Qian. *Chemical Biophysics: Quantitative Analysis of Cellular Systems*. Cambridge University Press, July 2008. ISBN 0521870704.

[12] A. Becskei and L. Serrano. Engineering stability in gene networks by autoregulation. *Nature*, 405(6786):590–593, June 2000. ISSN 0028-0836. doi: 10.1038/35014651.

[13] A. Becskei, B. B. Kaufmann, and A. van Oudenaarden. Contributions of low molecule number and chromosomal positioning to stochastic gene expression. *Nat. Genet.*, 37(9):937–944, Sept. 2005. ISSN 1061-4036. doi: 10.1038/ng1616.

[14] O. G. Berg, J. Paulsson, and M. Ehrenberg. Fluctuations and quality of control in biological cells: Zero-order ultrasensitivity reinvestigated. *Biophys. J.*, 79(3):1228–1236, 2000. doi: 10.1016/S0006-3495(00)76377-6.

[15] M. Bernaschi, F. Castiglione, A. Ferranti, C. Gavrila, M. Tinti, and G. Cesareni. ProtNet: A tool for stochastic simulations of protein interaction networks dynamics. *BMC Bioinformatics*, 8(Suppl 1):S4, 2007. doi: 10.1186/1471-2105-8-S1-S4.

[16] C. Blomberg. Fluctuations for good and bad: The role of noise in living systems. *Physics of Life Reviews*, 3:133–161, 2006. doi: 10.1016/j.plrev.2006.06.001.

[17] L. Boulianne, S. A. Assaad, M. Dumontier, and W. Gross. GridCell: A stochastic particle-based biological system simulator. *BMC Systems Biology*, 2(1):66, 2008. doi: 10.1186/1752-0509-2-66.

[18] H. Breuer and F. Petruccione. *The Theory of Open Quantum Systems*. Oxford University Press, 2002.

[19] G. Briggs and J. Haldane. A note on the kinetics of enzyme action. *Biochem. J.*, 19:338–339, 1925.

[20] K. Burrage, S. Mac, and T. Tian. Accelerated leap methods for simulating discrete stochastic chemical kinetics. In *Positive Systems*, pages 359–366. Springer Berlin / Heidelberg, 2006.

[21] X. Cai and Z. Xu. K-leap method for accelerating stochastic simulation of coupled chemical reactions. *J. Chem. Phys.*, 126(7):074102, Feb. 2007. doi: 10.1063/1.2436869.

[22] Y. Cao, D. Gillespie, and L. Petzold. Efficient step size selection for the tau-leaping simulation method. *J. Chem. Phys.*, 124:044109, 2006. doi: 10.1063/1.2159468.

[23] Y. Cao, D. T. Gillespie, and L. R. Petzold. Adaptive explicit-implicit tau-leaping method with automatic tau selection. *The Journal of Chemical Physics*, 126(22):224101, 2007. doi: 10.1063/1.2745299.

[24] B. Chen and Y. Wang. On the attenuation and amplification of molecular noise in genetic regulatory networks. *BMC Bioinformatics*, 7:52, Feb. 2006. doi: 10.1186/1471-2105-7-52.

[25] A. Cornish-Bowden. *Fundamentals of Enzyme Kinetics*. Portland Press, London, third edition, 2004.

[26] A. Csikász-Nagy, D. Battogtokh, K. C. Chen, B. Novák, and J. J. Tyson. Analysis of a generic model of eukaryotic cell cycle regulation. *Biophys. J.*, 90:4361–4379, 2006. doi: 10.1529/biophysj.106.081240.

[27] M. Delbrück. Statistical fluctuations in autocatalytic reactions. *The Journal of Chemical Physics*, 8(1):120–124, 1940. doi: 10.1063/1.1750549.

[28] M. Dogterom and S. Leibler. Physical aspects of the growth and regulation of microtubule structures. *Phys. Rev. Lett.*, 70(9):1347, Mar. 1993. doi: 10.1103/PhysRevLett.70.1347.

[29] E. R. Dougherty and M. L. Bittner. Causality, randomness, intelligibility, and the epistemology of the cell. *Curr. Genomics*, 11(17):221–23, June 2010. doi: doi:10.2174/138920210791233072.

[30] Y. Dublanche, K. Michalodimitrakis, N. Kümmerer, M. Foglierini, and L. Serrano. Noise in transcription negative feedback loops: Simulation and experimental analysis. *Molecular Systems Biology*, 2:41, 2006. doi: 10.1038/msb4100081.

[31] M. J. Dunlop, R. S. Cox, J. H. Levine, R. M. Murray, and M. B. Elowitz. Regulatory activity revealed by dynamic correlations in gene expression noise. *Nat. Genet.*, 40(12):1493–1498, Dec. 2008. ISSN 1061-4036. doi: 10.1038/ng.281.

[32] A. Eldar and M. B. Elowitz. Functional roles for noise in genetic circuits. *Nature*, 467(7312):167–173, Sept. 2010. ISSN 0028-0836. doi: 10.1038/nature09326.

[33] J. Elf. *Intracellular Flows and Fluctuations*. PhD thesis, Uppsala University, Teknisk-naturvetenskapliga vetenskapsområdet, Biology, Department of Cell and Molecular Biology, Sept. 2004. URL http://urn.kb.se/resolve?urn=urn:nbn:se:uu:diva-4291.

[34] J. Elf and M. Ehrenberg. Fast evaluation of fluctuations in biochemical networks with the linear noise approximation. *Genome Res.*, 13(11): 2475–2484, Nov. 2003. doi: 10.1101/gr.1196503.

[35] J. Elf and M. Ehrenberg. Spontaneous separation of bi-stable biochemical systems into spatial domains of opposite phases. *Syst. Biol.*, 1(2): 230–236, Dec. 2004. doi: 10.1049/sb:20045021.

[36] J. Elf, P. Lötstedt, and P. Sjöberg. Problems of high dimension in molecular biology. In *Proceedings of the 19th GAMM Seminar on High-dimensional problems*, pages 21–30. Max Planck Institute for Mathematics in the Sciences, Leipzig, Germany, Jan. 2003. URL http://www.mis.mpg.de/conferences/gamm/2003/.

[37] J. Elf, J. Paulsson, O. G. Berg, and M. Ehrenberg. Near-critical phenomena in intracellular metabolite pools. *Biophys. J.*, 84(1):154–170, Jan. 2003. doi: 10.1016/S0006-3495(03)74839-5.

[38] R. Ellis and A. Minton. Cell biology: Join the crowd. *Nature*, 425(6953): 27–28, Sept. 2003. doi: 10.1038/425027a.

[39] M. Elowitz, A. Levine, E. Siggia, and P. Swain. Stochastic gene expression in a single cell. *Science*, 297(5584):1183–1186, Aug. 2002. ISSN 1095-9203. doi: 10.1126/science.1070919.

[40] M. B. Elowitz and S. Leibler. A synthetic oscillatory network of transcriptional regulators. *Nature*, 403(6767):335–338, Jan. 2000. ISSN 0028-0836. doi: 10.1038/35002125.

[41] J. M. Epstein. Why model? *Journal of Artificial Societies and Social Simulation*, 11:4, 2008. URL http://econpapers.repec.org/RePEc:jas:jasssj:2008-57-1.

[42] C. P. Fall, E. S. Marland, J. M. Wagner, and J. J. Tyson. *Computational Cell Biology*. Springer, Berlin, February 2002. ISBN 0387953698. doi: 10.1007/b97701.

[43] D. Fell. *Understanding the Control of Metabolism*. Portland Press, London, 1997.

[44] L. Ferm, P. Lötstedt, and A. Hellander. A Hierarchy of Approximations of the Master Equation Scaled by a Size Parameter. Technical Report 2007-011, Uppsala University, Department of Information Technology, Apr. 2007. URL http://www.it.uu.se/research/publications/reports/2007-011/.

[45] J. E. Ferrell and W. Xiong. Bistability in cell signaling: How to make continuous processes discontinuous, and reversible processes irreversible. *Chaos*, 11(1):227–236, Mar. 2001. ISSN 1054-1500. doi: 10.1063/1.1349894.

[46] H. B. Fraser, A. E. Hirsh, G. Giaever, J. Kumm, and M. B. Eisen. Noise minimization in eukaryotic gene expression. *PLoS Biol.*, 2(6):e137, June 2004. doi: 10.1371/journal.pbio.0020137.

[47] C. Gardiner. *Handbook of Stochastic Models*. Springer, third edition, 2004.

[48] T. S. Gardner, C. R. Cantor, and J. J. Collins. Construction of a genetic toggle switch in *Escherichia coli. Nature*, 403(6767):339–342, Jan. 2000. ISSN 0028-0836. doi: 10.1038/35002131.

[49] C. v. Gend and U. Kummer. STODE—automatic stochastic simulation of systems described by differential equations. In Yi and Hucka, editors, *Proceedings of the 2nd International Conference on Systems Biology*, pages 326–333, Pasadena, 2001. Omnipress.

[50] M. Gibson and J. Bruck. Efficient exact stochastic simulation of chemical systems with many species and many channels. *J. Phys. Chem. A*, 104: 1876–188, 2000. doi: 10.1021/jp993732q.

[51] D. Gillespie. A general method for numerically simulating the stochastic time evolution of coupled chemical reactions. *Journal of Computational Physics*, 22:403–434, 1976. doi: 10.1016/0021-9991(76)90041-3.

[52] D. Gillespie. Exact stochastic simulation of coupled chemical reactions. *The Journal of Physical Chemistry*, 81(25):2340–2361, 1977. doi: 10.1021/jp993732q.

[53] D. Gillespie. A rigorous derivation of the chemical master equation. *Physica A*, 188:404–425, 1992. doi: 10.1016/0378-4371(92)90283-V.

[54] D. Gillespie. *Markov Processes*. Academic Press, 1992.

[55] D. Gillespie. The multivariate Langevin and Fokker–Planck equations. *American Journal of Physics*, 64(10):1246–1257, 1996. doi: 10.1119/1.18387.

[56] D. Gillespie. The chemical Langevin equation. *J. Chem. Phys.*, 113(1): 297–306, 2000. doi: 10.1063/1.481811.

[57] D. Gillespie. Approximate accelerated stochastic simulation of chemically reacting system. *J. Chem. Phys.*, 115(4):1716, 2001. doi: 10.1063/1.1378322.

[58] C. A. Gómez-Uribe and G. C. Verghese. Mass fluctuation kinetics: Capturing stochastic effects in systems of chemical reactions through coupled mean-variance computations. *J. Chem. Phys.*, 126(2):024109, Jan. 2007. doi: 10.1063/1.2408422.

[59] A. Goldbeter. *Biochemical Oscillations and Cellular Rythms*. Cambridge University Press, 1996.

[60] I. Golding and E. Cox. Physical nature of bacterial cytoplasm. *Phys. Rev. Lett.*, 96(9):098102, Mar. 2006. doi: 10.1103/PhysRevLett.96.098102.

[61] J. Goutsias. A hidden Markov model for transcriptional regulation in single cells. *IEEE/ACM Trans. Comput. Biol. Bioinform.*, 3(1):57–71, 2006. doi: 10.1109/TCBB.2006.2.

[62] J. Goutsias. Classical versus stochastic kinetics modeling of biochemical reaction systems. *Biophys. J.*, 92(7):2350–2365, Apr. 2007. doi: 10. 1529/biophysj.106.093781.

[63] R. Grima and S. Schnell. A mesoscopic simulation approach for modeling intracellular reactions. *Journal of Statistical Physics*, 128:139–164, July 2006. doi: 10.1007/s10955-006-9202-z.

[64] J. Hasty, J. Pradines, M. Dolnik, and J. Collins. Noise-based switches and amplifiers for gene expression. *Proc. Natl. Acad. Sci. U.S.A.*, 97(5): 2075–2080, 2000. doi: 10.1073/pnas.040411297.

[65] F. Hayot and C. Jayaprakash. The linear noise approximation for molecular fluctuations within cells. *Physical Biology*, 1(4):205–210, Dec. 2004. ISSN 1478-3975. doi: 10.1088/1478-3967/1/4/002.

[66] R. Heinrich and S. Schuster. *The Regulation of Cellular Systems*. Chapman and Hall, New York, 1996.

[67] R. H. Hering. Oscillations in Lotka-Volterra systems of chemical reactions. *J. Math. Chem.*, 5(2):197–202, 1990. ISSN 0259-9791. doi: 10.1007/BF01166429.

[68] D. J. Higham. An algorithmic introduction to numerical simulation of stochastic differential equations. *SIAM Review*, 43(3):525–546, 2001. doi: 10.1137/S0036144500378302.

[69] G. Hornung and N. Barkai. Noise propagation and signaling sensitivity in biological networks: A role for positive feedback. *PLoS Comput. Biol.*, 4(1):e8, Jan 2008. doi: 10.1371/journal.pcbi.0040008.

[70] W. Horsthemke and R. Lefever. *Noise-Induced Transitions*, volume 15 of *Springer Series in Synergetics*. Springer, 2006.

[71] J. House. *Principles of Chemical Kinetics*. Academic Press, San Diego, California, 2007.

[72] S. Intep, D. J. Higham, and X. Mao. Switching and diffusion models for gene regulation networks. *Multiscale Modeling & Simulation*, 8(1): 30–45, 2009. doi: 10.1137/080735412.

[73] T. Jahnke and W. Huisinga. Solving the chemical master equation for monomolecular reaction systems analytically. *J. Math. Biol.*, 54(1):1–26, Jan. 2007. doi: 10.1007/s00285-006-0034-x.

[74] M. Kaern, T. C. Elston, W. J. Blake, and J. J. Collins. Stochasticity in gene expression: from theories to phenotypes. *Nat. Rev. Genet.*, 6(6): 451–464, June 2005. ISSN 1471-0056. doi: 10.1038/nrg1615.

[75] N. v. Kampen. *Stochastic Processes in Physics and Chemistry*. Elsevier Amsterdam, Amsterdam, 3rd edition, 2007. ISBN 978-0-444-52965-7. URL http://www.sciencedirect.com/science/book/9780444529657.

[76] N. v. Kampen. The Langevin approach. In *Stochastic Processes in Physics and Chemistry*, pages 219–243. Elsevier, Amsterdam, 3rd edition, 2007. doi: 10.1016/B978-044452965-7/50012-X.

[77] N. v. Kampen. The Fokker–Planck equation. In *Stochastic Processes in Physics and Chemistry*, pages 193–218. Elsevier, Amsterdam, 3rd edition, 2007. doi: 10.1016/B978-044452965-7/50011-8.

[78] J. Keizer. *Statistical Thermodynamics of Nonequilibrium Processes*. Springer, Berlin, 1987.

[79] T. Kepler and T. Elston. Stochasticity in transcriptional regulation: origins, consequences, and mathematical representations. *Biophys. J.*, 81(6):3116–3136, Dec. 2001. doi: 10.1016/S0006-3495(01)75949-8.

[80] B. Kholodenko. Cell-signalling dynamics in time and space. *Nature Reviews: Molecular Cell Biology*, 7:165–176, 2006. doi: 10.1038/nrm1838.

[81] A. M. Kierzek. STOCKS: STOChastic Kinetic Simulations of biochemical systems with Gillespie algorithm. *Bioinformatics*, 18(3):470–481, Mar. 2002. doi: 10.1093/bioinformatics/18.3.470.

[82] M. Kimmel and D. E. Axelrod. *Branching Processes in Biology*, volume 19 of *Interdisciplinary Applied Mathematics*. Springer, 2002. doi: 10.1007/b97371.

[83] A. N. Kolmogorov. On the analytical methods of probability theory. *Math. Ann.*, 104:415–458, 1931.

[84] H. Kramers. Brownian motion in a field of force and the diffusion model of chemical reactions. *Physica*, 7(4):284–304, Apr. 1940. doi: 10.1016/S0031-8914(40)90098-2.

[85] K. Kruse and J. Elf. Kinetics in spatially extended systems in systems modeling in cellular biology. In Z. Szallasi, J. Stelling, and V. Periwal, editors, *System Modeling in Cellular Biology*, pages 177–198. The MIT Press, 2006.

[86] Y. Lan and G. A. Papoian. The interplay between discrete noise and nonlinear chemical kinetics in a signal amplification cascade. *J. Chem. Phys.*, 125(15):154901, Oct. 2006. doi: 10.1063/1.2358342.

[87] A. Lander. The edges of understanding. *BMC Biology*, 8(1):40, 2010. ISSN 1741-7007. URL http://www.biomedcentral.com/1741-7007/8/40.

[88] A. Lander. Pattern, Growth, and Control. *Cell*, 144(6):955–969, Mar. 2011. ISSN 0092-8674. doi: 10.1016/j.cell.2011.03.009.

[89] M. Lax. Fluctuations from the nonequilibrium steady state. *Reviews of Modern Physics*, 32(1):25–64, 1960. doi: 10.1103/RevModPhys.32.25.

[90] A. Leier, T. T. Marquez-Lago, and K. Burrage. Generalized binomial tau-leap method for biochemical kinetics incorporating both delay and intrinsic noise. *The Journal of Chemical Physics*, 128(20):205107, 2008. doi: 10.1063/1.2919124.

[91] J. Levine, H. Y. Kueh, and L. Mirny. Intrinsic fluctuations, robustness, and tunability in signaling cycles. *Biophys. J.*, 92(12):4473–4481, 2007. doi: 10.1529/biophysj.106.088856.

[92] Q. Li and X. Lang. Internal noise-sustained circadian rhythms in a drosophila model. *Biophys. J.*, 94(6):1983–1994, 2008. doi: 10.1529/biophysj.107.109611.

[93] T. Lipniacki, P. Paszek, A. Brasier, B. Luxon, and M. Kimmel. Transcriptional stochasticity in gene expression. *J. Theor. Biol.*, 238:348–367, 2006. doi: 10.1016/j.jtbi.2005.05.032.

[94] A. J. Lotka. Undamped oscillations derived from the law of mass action. *J. Am. Chem. Soc.*, 42(8):1595–1599, 1920. doi: 10.1021/ja01453a010.

[95] N. V. Mantzaris. From single-cell genetic architecture to cell population dynamics: Quantitatively decomposing the effects of different population heterogeneity sources for a genetic network with positive feedback architecture. *Biophys. J.*, 92(12):4271–288, Jun 2007. doi: 10.1529/biophysj.106.100271.

[96] MathWorks. Matlab. URL www.mathworks.com.

[97] H. H. McAdams and A. Arkin. Stochastic mechanisms in gene expression. *Proc. Natl. Acad. Sci. U.S.A.*, 94(3):814–819, 1997. doi: 10.1073/pnas.94.3.814.

[98] R. Metzler and J. Klafter. The random walk's guide to anomalous diffusion: A fractional dynamics approach. *Physical Reports*, 1:339, 2000. doi: 10.1016/S0370-1573(00)00070-3.

[99] L. Michaelis and M. Menten. Die Kinetik der Invertinwirkung. *Biochem. Z.*, 19:333–369, 1913.

[100] D. O. Morgan. *The Cell Cycle: Principles of Control.* Primers in Biology. New Science Press, 2007.

[101] Y. Morishita and K. Aihara. Noise-reduction through interaction in gene expression and biochemical reaction processes. *J. Theor. Biol.*, 228 (3):315–325, June 2004. doi: 10.1016/j.jtbi.2004.01.007.

[102] R. G. Mortimer. *Physical Chemistry.* Elsevier Academic Press, San Diego, California, 3rd edition, 2008.

[103] B. Munsky and M. Khammash. The finite state projection algorithm for the solution of the chemical master equation. *The Journal of Chemical Physics*, 124(4):044104, 2006. doi: 10.1063/1.2145882.

[104] B. Munsky and M. Khammash. A reduced model solution for the chemical master equation arising in stochastic analyses of biological networks. In *Decision and Control, 2006 45th IEEE Conference on,* pages 25–30, 2006.

[105] B. Novák and J. J. Tyson. Modelling the controls of the eukaryotic cell cycle. *Biochem. Soc. Trans.*, 31(6):1526–1529, 2003. URL http://www.biochemsoctrans.org/bst/031/bst0311526.htm.

[106] B. Novák and J. J. Tyson. Design principles of biochemical oscillators. *Nat. Rev. Mol. Cell Biol.*, 9(12):981–991, Dec. 2008. ISSN 1471-0072. doi: 10.1038/nrm2530.

[107] B. Novák, A. Csikasz-Nagy, B. Gyorffy, K. Chen, and J. J. Tyson. Mathematical model of the fission yeast cell cycle with checkpoint controls at the G1/S, G2/M and metaphase/anaphase transitions. *Biophys. Chem.*, 72(1-2):185–200, May 1998. doi: 10.1016/S0301-4622(98)00133-1.

[108] B. Novák, Z. Pataki, A. Ciliberto, and J. J. Tyson. Mathematical model of the cell division cycle of fission yeast. *Chaos*, 11(1):277–286, 2001. doi: 10.1063/1.1345725.

[109] B. Novák, K. Chen, and J. Tyson. Systems biology of the yeast cell cycle engine. In L. Alberghina and H. Westerhoff, editors, *Systems Biology*, volume 13 of *Topics in Current Genetics*, pages 305–324. Springer Berlin, 2005. doi: 10.1007/b137123.

[110] P. Nurse. A long twentieth century of the cell cycle and beyond. *Cell*, 100(1):71–78, Jan. 2000. doi: 10.1016/S0092-8674(00)81684-0.

[111] J. Osborne. Arguing to learn in science: The role of collaborative, critical discourse. *Science*, 328(5977):463–466, 2010. doi: 10.1126/science.1183944.

[112] E. M. Ozbudak, M. Thattai, H. N. Lim, B. I. Shraiman, and A. V. Oudenaarden. Multistability in the lactose utilization network of *Escherichia coli*. *Nature*, 427(6976):737–740, Feb. 2004. doi: 10.1038/nature02298.

[113] J. Pahle. *Stochastic Simulation and Analysis of Biochemical Networks*. PhD thesis, Humboldt-Universität zu Berlin, Fach Biophysik, 2008.

[114] A. Papoulis and S. U. Pillai. *Probability, Random Variables, and Stochastic Processes*. McGraw-Hill, fourth edition, 2001.

[115] P. Paszek. Modeling stochasticity in gene regulation: Characterization in the terms of the underlying distribution function. *Bull Math Biol*, 69:1567–1601, Mar. 2007. doi: 10.1007/s11538-006-9176-7.

[116] J. Paulsson. Summing up the noise. *Nature*, 427:415–418, 2004. doi: 10.1038/nature02257.

[117] J. Paulsson. Models of stochastic gene expression. *Phys. Life Rev.*, 2: 157–75, June 2005. doi: 10.1016/j.plrev.2005.03.003.

[118] J. Paulsson and M. Ehrenberg. Random signal fluctuations can reduce random fluctuations in regulated components of chemical regulatory networks. *Phys. Rev. Lett.*, 84:5447–5450, 2000. doi: 10.1103/PhysRevLett.84.5447.

[119] J. Paulsson and M. Ehrenberg. Noise in a minimal regulatory network: Plasmid copy number control. *Quarterly Reviews of Biophysics*, 34(1): 1–59, Feb. 2001. doi: 10.1017/S0033583501003663.

[120] J. Paulsson and J. Elf. Stochastic modeling of intracellular kinetics. In Z. Szallasi, J. Stelling, and V. Periwal, editors, *System Modeling in Cellular Biology*, pages 149–176. The MIT Press, Cambridge, Massachusetts, 2006.

[121] J. Paulsson, O. Berg, and M. Ehrenberg. Stochastic focusing: fluctuation-enhanced sensitivity of intracellular regulation. *Proc. Natl. Acad. Sci. U.S.A.*, 97:7148–7153, 2000. doi: 10.1073/pnas.110057697.

[122] J. M. Pedraza and A. van Oudenaarden. Noise propagation in gene networks. *Science*, 307(5717):1965–1969, 2005. doi: 10.1126/science. 1109090.

[123] S. Peles, B. Munsky, and M. Khammash. Reduction and solution of the chemical master equation using time scale separation and finite state projection. *J. Chem. Phys.*, 125(20):204104, Nov 2006. doi: 10.1063/1.2397685.

[124] J. Puchalka and A. M. Kierzek. Bridging the gap between stochastic and deterministic regimes in the kinetic simulations of the biochemical reaction networks. *Biophys. J.*, 86(3):1357–1372, Mar. 2004. doi: 10. 1016/S0006-3495(04)74207-1.

[125] H. Qian. From discrete protein kinetics to continuous Brownian dynamics: A new perspective. *Protein Sci.*, 11(1):1–5, 2002. doi: 10.1110/ps.18902.

[126] A. Raj and A. van Oudenaarden. Nature, nurture, or chance: Stochastic gene expression and its consequences. *Cell*, 135(2):216 – 226, 2008. ISSN 0092-8674. doi: 10.1016/j.cell.2008.09.050.

[127] C. V. Rao, D. M. Wolf, and A. P. Arkin. Control, exploitation and tolerance of intracellular noise. *Nature*, 420(6912):231–237, Nov. 2002. doi: 10.1038/nature01258.

[128] J. M. Raser and E. K. O'Shea. Noise in gene expression: Origins, consequences, and control. *Science*, 309(5743):2010–2013, Sept. 2005. doi: 10.1126/science.1105891.

[129] A. S. Ribeiro and J. Lloyd-Price. SGN Sim, a stochastic genetic networks simulator. *Bioinformatics*, 23(6):777–779, Mar. 2007. doi: 10.1093/ bioinformatics/btm004.

[130] A. S. Ribeiro, D. A. Charlebois, and J. Lloyd-Price. CellLine, a stochastic cell lineage simulator. *Bioinformatics*, 23(24):3409–3411, 2007. doi: 10.1093/bioinformatics/btm491.

[131] A. Rényi. On the theory of order statistics. *Acta Mathematica Hungarica*, 4(3):191–231, Sept. 1953. doi: 10.1007/BF02127580.

[132] H. E. Samad and M. Khammash. Intrinsic noise rejection in gene networks by regulation of stability. In *First International Symposium on Control, Communications and Signal Processing*, pages 187–190, Hammamet, Tunisia, 2004. doi: 10.1109/ISCCSP.2004.1296252.

[133] M. Samoilov, S. Plyasunov, and A. P. Arkin. Stochastic amplification and signaling in enzymatic futile cycles through noise-induced bistability with oscillations. *Proc. Natl. Acad. Sci. U.S.A.*, 102(7):2310–2315, Feb. 2005. doi: 10.1073/pnas.0406841102.

[134] G. Sanguinetti, A. Ruttor, M. Opper, and C. Archambeau. Switching regulatory models of cellular stress response. *Bioinformatics*, 25(10): 1280–1286, 2009. doi: 10.1093/bioinformatics/btp138.

[135] M. Saxton. Anomalous diffusion due to obstacles: A Monte Carlo study. *Biophys. J.*, 66:394–401, Feb. 1994. doi: 10.1016/S0006-3495(94) 80789-1.

[136] M. Saxton and K. Jacobson. Single-particle tracking: Applications to membrane dynamics. *Annu. Rev. Biophys. Biomol. Struct.*, 26:373–399, 1997. doi: 10.1146/annurev.biophys.26.1.373.

[137] F. Schlögl. Chemical reaction models for non-equilibrium phase transitions. *Zeitschrift für Physik A Hadrons and Nuclei*, 253(2):147–161, Apr. 1972. doi: 10.1007/BF01379769.

[138] S. Schnell and T. Turner. Reaction kinetics in intracellular environments with macromolecular crowding: Simulations and rate laws. *Prog. Biophys. Mol. Biol.*, 85(2-3):235–260, 2004. doi: 10.1016/j.pbiomolbio. 2004.01.012.

[139] M. Scott and B. Ingalls. Using the linear noise approximation to characterize molecular noise in reaction pathways. In *Proceedings of the AIChE Conference on Foundations of Systems Biology in Engineering (FOSBE)*, Santa Barbara, California, Aug. 2005.

[140] M. Scott, B. Ingalls, and M. Kaern. Estimations of intrinsic and extrinsic noise in models of nonlinear genetic networks. *Chaos*, 16(2):026107, June 2006. doi: 10.1063/1.2211787.

[141] V. Shahrezaei, J. F. Ollivier, and P. S. Swain. Colored extrinsic fluctuations and stochastic gene expression. *Mol. Syst. Biol.*, 4:196, May 2008. doi: 10.1038/msb.2008.31.

[142] T. Shibata and M. Ueda. Noise generation, amplification and propagation in chemotactic signaling systems of living cells. *Biosystems*, 93 (1-2):126–132, 2008. doi: 10.1016/j.biosystems.2008.04.003.

[143] K. Singer. Application of the theory of stochastic processes to the study of irreproducible chemical reactions and nucleation processes. *Journal of the Royal Statistical Society. Series B (Methodological)*, 15 (1):92–106, 1953. ISSN 0035-9246. URL http://www.jstor.org/pss/2983726.

[144] R. Steuer. Effects of stochasticity in models of the cell cycle: From quantized cycle times to noise-induced oscillations. *J. Theor. Biol.*, 228 (3):293–301, June 2004. doi: 10.1016/j.jtbi.2004.01.012.

[145] A. Sveiczer and B. Novák. Regularities and irregularities in the cell cycle of the fission yeast, Schizosaccharomyces pombe. *Acta Microbiologica et Immunologica Hungarica*, 49(2):289–304, May 2002. doi: 10.1556/AMicr.49.2002.2-3.17.

[146] A. Sveiczer, B. Novák, and J. Mitchison. The size control of fission yeast revisited. *J. Cell Sci.*, 109(12):2947–2957, 1996. URL http://jcs.biologists.org/cgi/content/abstract/109/12/2947.

[147] A. Sveiczer, A. Csikasz-Nagy, B. Gyorffy, J. J. Tyson, and B. Novák. Modeling the fission yeast cell cycle: Quantized cycle times in $wee1^{--}$ $cdc25Delta$ mutant cells. *Proc. Natl. Acad. Sci. U.S.A.*, 97(14):7865–7870, 2000. doi: 10.1073/pnas.97.14.7865.

[148] A. Sveiczer, J. J. Tyson, and B. Novák. A stochastic, molecular model of the fission yeast cell cycle: Role of the nucleocytoplasmic ratio in cycle time regulation. *Biophys. Chem.*, 92(1-2):1–15, Sept. 2001. doi: 10.1016/S0301-4622(01)00183-1.

[149] M. Tang. The mean and noise of stochastic gene transcription. *J. Theor. Biol.*, 253:271–280, 2008. doi: 10.1016/j.jtbi.2008.03.023.

[150] Y. Tao, Y. Jia, and T. G. Dewey. Stochastic fluctuations in gene expression far from equilibrium: Omega expansion and linear noise approximation. *J. Chem. Phys.*, 122(12):124108, Mar. 2005. doi: 10.1063/1.1870874.

[151] M. Thattai and A. van Oudenaarden. Attenuation of noise in ultrasensitive signaling cascades. *Biophys. J.*, 82(6):2943–2950, June 2002. doi: 10.1016/S0006-3495(02)75635-X.

[152] Y. Togashi and K. Kaneko. Switching dynamics in reaction networks induced by molecular discreteness. *Journal of Physics—Condensed Matter*, 19(6):065150, Feb. 2007. doi: 10.1088/0953-8984/19/6/065150.

[153] S. Toulmin. *The Uses of Argument*. Cambridge University Press, Cambridge, 1958.

[154] M. Turcotte, J. Garcia-Ojalvo, and G. M. Süel. A genetic timer through noise-induced stabilization of an unstable state. *Proc. Natl. Acad. Sci. U.S.A.*, 105(41):15732–15737, 2008. doi: 10.1073/pnas.0806349105.

[155] T. Turner, S. Schnell, and K. Burrage. Stochastic approaches for modelling in vivo reactions. *Comput. Biol. Chem.*, 28(3):165–178, July 2004. doi: 10.1016/j.compbiolchem.2004.05.001.

[156] J. J. Tyson, A. Csikasz-Nagy, and B. Novák. The dynamics of cell cycle regulation. *BioEssays*, 24(12):1095–1109, 2002. doi: 10.1002/bies.10191.

[157] J. J. Tyson, K. C. Chen, and B. Novák. Sniffers, buzzers, toggles and blinkers: Dynamics of regulatory and signaling pathways in the cell. *Curr. Opin. Cell Biol.*, 15(2):221–231, Apr. 2003. doi: 10.1016/S0955-0674(03)00017-6.

[158] M. Ullah and O. Wolkenhauer. Family tree of Markov models in systems biology. *IET Systems Biology*, 1(4):247–254, 2007. doi: 10.1049/iet-syb:20070017.

[159] M. Ullah and O. Wolkenhauer. Investigating the two-moment characterisation of subcellular biochemical networks. *J. Theor. Biol.*, 260(3):340–352, Oct. 2009. ISSN 0022-5193. doi: 10.1016/j.jtbi.2009.05.022.

[160] T. Ushikubo, W. Inoue, M. Yoda, and M. Sasai. Testing the transition state theory in stochastic dynamics of a genetic switch. *Chem. Phys. Lett.*, 430(1-3):139–143, Oct. 2006. doi: 10.1016/j.cplett.2006.08.114.

[161] J. Vera, E. Balsa-Canto, P. Wellstead, J. R. Banga, and O. Wolkenhauer. Power-law models of signal transduction pathways. *Cell Signalling*, 19(7):1531–1541, July 2007. doi: 10.1016/j.cellsig.2007.01.029.

[162] J. M. G. Vilar, H. Y. Kueh, N. Barkai, and S. Leibler. Mechanisms of noise-resistance in genetic oscillators. *Proc. Natl. Acad. Sci. U.S.A.*, 99(9):5988–5992, Apr 2002. doi: 10.1073/pnas.092133899.

[163] E. O. Voit. *Computational Analysis of Biochemical Systems : A Practical Guide for Biochemists and Molecular Biologists*. Cambridge University Press, September 2000. doi: 10.2277/0521785790.

[164] V. Volterra. Fluctuations in the abundance of a species considered mathematically. *Nature*, 118:558–560, 1926. doi: 10.1038/119012b0.

[165] D. J. Wilkinson. *Stochastic Modelling for Systems Biology*. Mathematical & Computational Biology. Chapman & Hall/CRC, London, Apr. 2006. ISBN 1584885408.

[166] D. J. Wilkinson. Stochastic modelling for quantitative description of heterogeneous biological systems. *Nat. Rev. Genet.*, 10(2):122–133, Feb. 2009. ISSN 1471-0056. doi: 10.1038/nrg2509.

[167] V. Wolf, R. Goel, M. Mateescu, and T. Henzinger. Solving the chemical master equation using sliding windows. *BMC Systems Biology*, 4(1):42, 2010. ISSN 1752-0509. doi: 10.1186/1752-0509-4-42.

[168] O. Wolkenhauer and J.-H. S. Hofmeyr. An abstract cell model that describes the self-organization of cell function in living systems. *J. Theor. Biol.*, 246(3):461–476, June 2007. doi: 10.1016/j.jtbi.2007.01.005.

[169] O. Wolkenhauer, M. Ullah, W. Kolch, and K.-H. Cho. Modeling and simulation of intracellular dynamics: Choosing an appropriate framework. *IEEE Trans. Nanobioscience*, 3(3):200–207, Sept. 2004. doi: 10.1109/TNB.2004.833694.

[170] O. Wolkenhauer et al. Systems biologists seek fuller integration of systems biology approaches in new cancer research programs. *Cancer Res.*, 70(1):12–13, 2010. doi: 10.1158/0008-5472.CAN-09-2676.

[171] M. R. Wright. *Introduction to Chemical Kinetics*. Wiley-Interscience, West Sussex, 2004.

[172] M. Yi, Y. Jia, J. Tang, X. Zhan, L. Yang, and Q. Liu. Theoretical study of mesoscopic stochastic mechanism and effects of finite size on cell cycle of fission yeast. *Physica A: Statistical Mechanics and its Applications*, 387(1):323–334, Jan. 2008. doi: 10.1016/j.physa.2007.07.018.

[173] M. Yoda, T. Ushikubo, W. Inoue, and M. Sasai. Roles of noise in single and coupled multiple genetic oscillators. *J. Chem. Phys.*, 126(11): 115101, Mar 2007. doi: 10.1063/1.2539037.

[174] J. Zamborszky, C. I. Hong, and A. Csikasz Nagy. Computational analysis of mammalian cell division gated by a circadian clock: Quantized cell

cycles and cell size control. *J. Biol. Rhythms*, 22(6):542–553, 2007. doi: 10.1177/0748730407307225.

[175] S. Zeiser, U. Franz, and V. Liebscher. Autocatalytic genetic networks modeled by piecewise-deterministic Markov processes. *Journal of Mathematical Biology*, 60(2):207–246, Feb. 2010. ISSN 0303-6812. doi: 10.1007/s00285-009-0264-9.

[176] Y. Zhang, H. Yu, M. Deng, and M. Qian. Nonequilibrium model for yeast cell cycle. In *Computational Intelligence and Bioinformatics*, pages 786–791. Springer Berlin, 2006. doi: 10.1007/11816102.

Glossary

The glossary provides brief descriptions of some technical terms used in the main text. In addition to printed encyclopedias and dictionaries (e.g., the *Cambridge Dictionary of Statistics*, 2nd ed, B.S. Everitt, Cambridge University Press, 2002), the Internet provides a number of resources along those lines. These include

Wikipedia: http://en.wikipedia.org

PlanetMath: http://planetmath.org

Wolfram MathWorld: http://mathworld.wolfram.com

Springer Encyclopedia of Mathematics: http://eom.springer.de/

Oxford Dictionary of Statistics: http://www.encyclopedia.com

IUPAC: http://goldbook.iupac.org/

2MA equations A closed system of ODEs for the dynamics of the mean and (co)variance of a continuous-time discrete-state Markov process that models a biochemical reaction system.

Active site Region of an enzyme surface to which a substrate molecule binds in order to undergo a catalyzed reaction.

Algebra A branch of mathematics that generalizes arithmetic operations with numbers to operations with variables, matrices, etc.

Amino acid Class of biochemical compounds from which proteins are composed. Around 20 amino acids are present in proteins.

Analysis A branch of mathematics concerned primarily with limits of functions, sequences, and series.

Analytic function A function possessing derivatives of all orders and agreeing with its Taylor expansion locally.

Analytical solution A solution to an equation or set of equations that can be explicitly written in terms of known functions and constants (i.e., in closed form).

Apoptosis The death of some of an organism's cells as part of its natural growth and development. Also called programmed cell death.

Argument The argument of a function is the element to which a function is applied; usually the independent variable of the function.

Arrival The occurrence of an event in a counting process.

Association Any process of combination (especially in solution) that depends on relatively weak chemical bonding.

ATP The principal carrier of chemical energy in cells.

Attractor A region of the space describing the temporal solution of a dynamic system toward which trajectories nearby converge, that is, are attracted to. An attractor can be an equilibrium point or a circle. An attracting region that has no individual equilibrium point or cycle is referred to as a chaotic or strange attractor.

Autocatalysis Reaction catalyzed by one of its products, creating a positive feedback (self-amplifying) effect on the reaction rate.

Autoinhibition Mechanism for inhibiting a system's own activity; e.g., Raf contains an autoregulatory domain that inhibits its own activity by binding to its catalytic domain. The autoregulatory domain is relieved from the catalytic domain by phosphorylation of characteristic residues.

Autonomous A system (of differential equations) is said to be autonomous if it does not explicitly depend on time.

Avogadro's number The number of molecules in a mole of a substance, approximately 6.0225×10^{23}.

Bifurcation point An instability point in which a single equilibrium condition is split into two. At a bifurcation point the dynamics of a system change structurally.

Binding site A specific stretch of sequence (protein, DNA or RNA) to which another molecule or entity is able to bind and form a chemical bond.

Binomial distribution The discrete probability distribution of the number of successes in a sequence of independent yes/no experiments, each of which yields success with a fixed probability.

Bioinformatics The management and analysis of genomic data, most commonly using tools and techniques from computer science.

Biological activity The potential action that a biological entity has on other entities. Examples are enzymatic activity and binding activity.

Birth–death process A method for describing the size of a population in which the population increases or decreases by one unit or remains constant over short time periods.

Bistability A property of of system (of equations) for which two stable solutions exist. The existence of multiple stable solutions is known as multistability.

Black-box model A model that aims to determine the functional relationship between known system input and output when the specifics of the system structure are unknown. Black-box models are often built from existing data using some form of regression analysis.

Borel set Any set in a topological space that can be formed from open sets (or, equivalently, from closed sets) through the operations of countable union, countable intersection, and relative complement.

Brownian motion The erratic motion, visible through a microscope, of small grains suspended in a fluid.

Catalytic site A catalytic site is the region that confers specificity of a substrate for the binding entity, and where specific reactions take place in the conversion of the substrate to the product.

Cell biology An academic discipline that studies cells—their physiological properties, their structure, the organelles they contain, interactions with their environment, their life cycle, division, and death.

Cell fate decision The process whereby cells assume different, functionally important and heritable fates without an associated genetic or environmental difference.

Cell differentiation The series of events involved in the development of a specialized cell having specific structural, functional, and biochemical properties.

Cell proliferation Process of cell growth and cell division.

Channel Participating entity that allows another participating entity to pass through it, possibly connecting different compartments.

Chapman–Kolmogorov equation An identity relating the joint probability distributions of different sets of coordinates on a stochastic process.

Chemical Langevin equation A stochastic differential equation describing the time evolution of a subset of the degrees of freedom.

Chemical kinetics A branch of physical chemistry concerned with the mechanisms and rates of chemical reactions.

Chemical reaction A change in which a substance (or substances) is changed into one or more new substances.

Chemical species Atoms, molecules, molecular fragments, ions, etc., being subjected to a chemical process or to a measurement.

Closed form An expression or solution in terms of well-understood quantities.

Coefficient A numerical or constant multiplier of a variable in an algebraic term.

Coefficient of variation A statistical measure of the dispersion of data points in a data series around the mean.

Collision theory Theory of chemical reaction proposing that the rate of product formation is equal to the number of reactant–molecule collisions multiplied by a factor that corrects for low-energy-level collisions.

Complementary cumulative distribution function The probability for a realization of the variable to be larger than a given value.

Complex Consisting of interconnected or interwoven parts.

Conditional probability The probability that an event will occur, given that one or more other events have occurred.

Conformational change Biochemical reaction that does not result in the modification of covalent bonds of reactants, but rather modifies the conformation of some reactants, that is, the relative position of their atoms in space.

Continuous function A function for which the value changes gradually.

Continuous random variable A random variable for which the data can take infinitely many values.

Continuous-time Markov process A stochastic process that satisfies the Markov property and takes values from a set called the state space.

Control Target or set-point tracking, making the system sensitive to changes in the input. See also regulation and homeostasis.

Conservation Mathematical expression stating that a quantity is conserved in a system, whatever happens within the boundaries of that system.

Consistency condition The requirement that a mathematical theory be free from contradiction.

Conversion Biochemical reaction that results in the modification of some covalent bonds.

Copy number The number of plasmid or other DNA molecules in a cell.

Counting process When events occur in a sequence, the stochastic process defined by the number of events (for some definition of an event) that have occurred up to a given time is a counting process.

Covalence The number of electron pairs an atom can share with other atoms.

Covalent modification Alteration in the structure of a macromolecule by enzymatic means, resulting in a change in the properties of that macromolecule; frequently, this type of modification is physiologically relevant.

Covariance The covariance of a pair of random variables is the expectation of products of their deviations from their respective means.

Cumulative distribution function The probability for a realization of the variable to be less than a given value.

Cytokine Extracellular signal protein or peptide that acts as a local short-distance mediator in cell–cell communication. Cytokines are called lymphokines if produced by lymphocytes, interleukines if produced by leucocytes, and monokines if produced by monocytes and macrophages.

Cytoplasm Contents of a cell that are contained within its plasma membrane but, in the case of eukaryotic cells, outside the nucleus.

Damped oscillations An oscillation in which the amplitude decreases over time.

Degradation The change of a chemical species into a form that is less complex/interesting/useful. The latter qualification depends on the context. Degradation can be thought of as an increase in entropy.

Deterministic model A model in which future states are fully determined by the past and present states, frequently built using differential equations.

Differentiation process by which the cell acquires specialized functional properties.

Differentiable A system (usually a process described by differential equations) is called differentiable if its phase space has the structure of a differentiable manifold, and the change of state is described by differentiable functions.

Differential equation A differential equation gives a relation between the value of some varying quantity and the rate at which that quantity is changing, and perhaps the rate at which that rate is changing, and so on. The numerical solution of a differential equation is a table of values of the varying quantity that to a good approximation satisfy both the differential equation and some given conditions on the initial values of this quantity and of its rates of change.

Differential–difference equation Differential in time and difference in states.

Dimer A protein molecule that consists of two subunits (monomers) held together by some type of bonding; homodimer: the subunits are identical; heterodimer: the subunits are different.

Dimerization The process by which two molecules of the same chemical composition form a condensation product or polymer.

Discrete random variable A variable that can assume only a countable number of values and the sum of the probabilities is one.

Dissipation Wasteful expenditure or consumption.

Dissociation The temporary or reversible process in which a molecule or ion is broken down into smaller molecules or ions.

DNA replication Process in which a DNA duplex is transformed into two identical DNA duplexes.

Dynamic system A system that changes with time.

Elementary event A single point of a sample space.

Emergent properties Properties of a system that arise from the interactions among its components that cannot be deduced from their individual behavior.

Empirical model A nonmechanistic model that shows good agreement with existing experimental data and can be used to predict outcomes in separate but similar data sets.

Ensemble A unit or group of complementary parts that contribute to a single effect.

Enzyme Protein that catalyzes a specific chemical reaction.

Equilibrium State in which there is no net change in a system. In a chemical reaction the equilibrium is defined by the state at which the forward and reverse rates are equal.

Equilibrium point Point such that the derivatives of a system of differential equations are zero. An equilibrium point may be stable (then called an attractor) or unstable (repellor).

Expectation The expected value of a random variable.

Expression Production of a protein that has directly observable consequences.

Extinction The condition of being extinguished; complete depletion/disappearance of a chemical species.

External noise Noise that arises in (the behavior of) a system from interaction with other systems such as cell–cell interaction.

Extrinsic noise See external noise.

Event Something that happens or is regarded as happening; an occurrence, especially one of some importance; the outcome, issue, or result of anything.

Fano factor A measure of the dispersion of a probability distribution of a Fano noise.

Feedback inhibition Regulatory mechanism in metabolic pathways: an enzyme further up in the pathway is inhibited by a product further down in that pathway.

Feedback loop A loop structure in which the output signal y produced by an element upon receiving an input signal u is also an input signal to the element generating signal u producing a down-regulation/up-regulation of signal u.

Feedforward loop A loop structure in which two signals generated by a system element converge on an element downstream from this origin. Feedforward control can either speed up a system's dynamics or destabilize it.

Finite-dimensional A process is called finite-dimensional if its phase space is finite-dimensional, i.e., if the number of parameters needed to describe its states is finite.

Fission yeast A species of yeast.

Fixed point See steady state.

Formal system A mathematical framework in which to represent natural systems.

Function A relation between two sets that describes unique associations among the elements of the two sets. A function is sometimes called a mapping or transformation.

Function handle A Matlab object, created inline or from a function, that can be used as a function.

Gaussian process A stochastic process for which any finite linear combination of samples will be normally distributed (or, more generally, any linear functional applied to the sample function will give a normally distributed result).

Gene A region of genomic sequence, corresponding to a unit of inheritance, that is associated with regulatory region, transcribed regions and/or other functional sequence regions.

Gene product The macromolecules, RNA or proteins, that are the result of gene expression.

Gene expression The process by which the information coded in the genome is transcribed into RNA. Expressed genes include those for which the RNA is not translated into proteins.

Gene transcription The process by which genetic information is copied from DNA to RNA, resulting in a specific protein formation.

Genome The entirety of genetic material (DNA) of a cell or an organism.

Gillespie algorithm Generates a statistically correct trajectory (possible solution) of a stochastic equation.

Growth factor Extracellular signaling molecule that can stimulate a cell to grow or proliferate.

G-proteins Small monomeric GTP-binding proteins (e.g., Ras), molecular switches that modulate the connectivity of a signaling cascade: resting G-proteins are loaded with GDP and inactive; replacement of GDP with GTP by exchange factors means activation.

Heterodimer A protein made of paired polypeptides that differ in their amino acid sequences.

Heterodimerization An association of two identical molecules linked together.

Heterogeneity The quality or state of being heterogeneous.

Homeostasis Regulation to maintain the level of a variable. See also regulation.

Homogeneous process A stochastic process is said to be homogeneous in space if the transition probability between any two state values at two given times depends only on the difference between those state values. The process is homogeneous in time if the transition probability between two given state values at any two times depends only on the difference between those times.

Identifiability A property that a model must satisfy in order for inference to be possible. We say that the model is identifiable if it is theoretically possible to learn the true value of the model's underlying parameter after obtaining an infinite number of observations from it.

Independent events Such events that the occurrence of one of them makes it neither more nor less probable that the other occurs.

Independent increments A stochastic process is said to have independent increments when its increments during nonoverlapping time intervals are independent random variables.

Infinitesimal Infinitely small. Infinitesimal quantities are used to define integrals and derivatives, and are studied in the branch of mathematics called analysis.

Inhibition Negative modulation of the execution of a process.

in silico A term used in reference to systems created, solved, or simulated using a computer.

Intermediate state A state through which a system may pass during transition from an initial state to a final state.

Internal noise Noise that arises (in the behavior of) a system by the very discrete nature of events (such as chemical reactions) happening inside the system.

Intrinsic noise See internal noise.

in vitro Experimental procedures taking place in an isolated cell-free extract. Cells growing in culture, as opposed to an organism.

in vivo In an intact cell or organism.

Ion channel Ion channels are pore-forming proteins that help establish and control a small voltage gradient across the plasma membrane in living cells by allowing the flow of ions along their electrochemical gradient.

Irreversible Impossible to reverse or be reversed.

Isoforms Closely homologous proteins (from different genes) that perform similar or only slightly different functions, e.g., under tissue-specific control. Two or more RNAs that are produced from the same gene by different transcription and/or differential RNA splicing are referred to as isoforms.

Isogenic cells Cells having the same genetic makeup; characterized by identical genetic composition; being genetically alike.

Isomerization Process of transforming one chemical molecule into another without changing its composition. The principal reactant and principal product are isomers of each other.

Jump A process is said to have taken a jump when it has moved discontinuously or has changed by a large amount during a short period.

Jump process A type of stochastic process that has large discrete movements (jumps), rather than small continuous movements.

Kinase Enzyme that catalyzes the phosphorylation of a protein.

Langevin equation A stochastic differential equation describing the time evolution of a subset of the degrees of freedom.

Ligand A physical entity (such as a molecule) that binds to a site on a receptor's surface by intermolecular forces.

Linear noise approximation Analytic supplement to numerical simulations. The approximation proceeds in two steps. First, the fluctuations are separated from the macroscopic trajectory of the system. Second, the discrete state space is smeared into a continuum. The master equation so modified is then expanded in inverse powers of the system size.

Linear operator A function between two vector spaces that preserves the operations of vector addition and scalar multiplication.

Linear system A system is nonlinear if changes in the output are not proportional to changes in the input.

Linearization Taylor expansion of a dynamical system in the dependent variable about a specific solution, discarding all but the terms linear in the dependent variable.

Locus The position of a gene on a chromosome, the DNA of that position; usually restricted to the main regions of DNA that are expressed.

M-phase The division of the mother cell into two daughter cells, genetically identical to each other and to their parent cell.

MAP-kinase Mitogen-activated protein kinase that performs a crucial step in transmitting signals from the plasma membrane to the nucleus.

Markov process A Markov process is a mathematical model for the random evolution of a memoryless system, that describes the likelihood of a given future state, at any given point in time, depending only on its present state, and not on any past states. See also Markov property, Markov chain.

Markov property A stochastic process is said to have the Markov property, or to be memoryless, if its future and past are independent and conditional only on the present state of the system.

Markov chain The term Markov chain is synonymous to a Markov process with a discrete (finite or countable) state space. Usually, a Markov chain is defined for a discrete set of times although some authors use the same terminology where "time" can take continuous values.

Mass action Mathematical model describing the behavior of solutions in dynamic equilibrium.

Master equation In physics, a master equation is a phenomenological set of first-order differential equations describing the time evolution of the probability of a system to occupy each one of a discrete set of states.

Mean The (statistical) mean, or average, of a random quantity is a a value one expects in most of the observations made on the quantity.

Measure In mathematics, more specifically in measure theory, a measure on a set is a systematic way to assign to each suitable subset a number, intuitively interpreted as the size of the subset. In this sense, a measure is a generalization of the concepts of length, area, volume, etc.

Mechanistic model A model that describes the physical processes that give rise to observed properties of the system. Variables and parameters of the system correspond to physical quantities and rates that can be measured empirically.

Memoryless property A property of an exponential random variable such that the future is independent of the past i.e., the fact that it has not happened yet, tells us nothing about how much longer it will take before it does happen.

Mesoscopic Pertaining to a size regime, intermediate between the microscopic and the macroscopic, that is characteristic of a region where a large number of particles can interact in a quantum-mechanically correlated fashion.

Metabolic Control Analysis (MCA) Method for analyzing variation in fluxes and intermediate concentrations in a metabolic pathway relating to the effects of the different enzymes that constitute the pathway and external parameters. The building blocks of MCA are control coefficients, elasticity coefficients, and response coefficients.

Metabolism Metabolism is the set of chemical reactions that happen in living organisms to maintain life. These processes allow organisms to grow and reproduce, maintain their structures, and respond to their environments. Metabolism is usually divided into two categories. Catabolism breaks down organic matter, for example to harvest energy in cellular respiration. Anabolism uses energy to construct components of cells such as proteins and nucleic acids.

Metabolite Substance produced by metabolism or by a metabolic process.

Michaelis–Menten kinetics Mathematical model describing enzyme kinetic reaction in a good approximation.

Mitogen Substance that stimulates cell mitosis.

Mitosis The entire process of cell division including division of the nucleus and the cytoplasm.

Mixed-type random variable A random variable whose distribution function has a jump discontinuity at a countable number of points and that increases continuously at least at one interval of states.

mRNA RNA resulting from the transcription of DNA that is used for protein synthesis.

Modeling framework Set of assumptions that underlie a mathematical description.

Molar concentration A measure of the concentration of a solute in a solution, or of any molecular, ionic, or atomic species in a given volume.

Molecular noise Stochastic fluctuations in molecular expression levels originating from the inherent indeterminism of molecular processes and the unpredictable variability of the extracellular environment.

Monomer A single subunit of a protein molecule.

Moments The expected value of a positive integral power of a random variable. The first moment is the mean of the distribution.

Monostable Having only one stable state.

Monte Carlo simulation A widespread method used to obtain observable quantities that depend on random variables whose probability distributions are known. Monte Carlo methods can be used to introduce stochasticity into a model but are also used to sample the parameter space of deterministic models.

Multistable system A dynamical system that supports the existence of two or more coexisting attractors for some region of parameter space.

Natural system An aspect of the phenomenal world, studied in the natural sciences.

Negative feedback Feedback that reduces the output of a system, such as the action of heat on a thermostat to limit the output of a furnace or the accumulation of toxic waste products by a growing population of bacteria.

Network A (metabolic/signaling) network is the complete set of metabolic/signaling and physical processes that determine the physiological and biochemical properties of a cell. As such, these networks comprise the chemical reactions of metabolism/signaling as well as the regulatory interactions that guide these reactions.

Noise Randomness, uncertainty, or unpredictability of the behavior of a system. Noise is either an inherent part of a system (which is referred to as intrinsic or internal noise) or is an influence of the environment (which is referred to as extrinsic or external noise).

Nonlinearity Linearity is defined in terms of functions that have the property $f(x + y) = f(x) + f(y)$ and $f(ax) = af(x)$. This means that the result f may not be proportional to the input x or y.

Nontrivial solution A solution of a set of homogeneous linear equations in which at least one of the variables has a value different from zero.

Nuclear pore Nuclear pores are large protein complexes that cross the inner—nuclear membrane in eukaryotic cells.

Numerical solution Computationally determined solution to an equation or system of equations, typically necessary when an analytical solution is intractable. A numerical solution is an approximation of the closed-form solution, but it can be calculated to any desired level of precision, given enough time and computational power.

Observable An entity that can be measured quantitatively.

Open system A system in which there is a continual exchange of material, energy, and information with the environment.

Orbit The set of points in phase space through which a trajectory passes.

Organization Pattern or configuration of processes.

Oscillation The periodic variation, typically in time, of some measure as seen, for example, in a swinging pendulum.

Partition (of a set) A collection of nonempty subsets such that every element belongs to one and only one of the subsets.

Pathway The term pathway refers to a series of chemical reactions occurring within a cell. Collections of pathways are grouped into the metabolic network and signaling network.

Peptide A small chain of amino acids linked by peptide bonds.

Phase space Phase space is the collection of possible states of a dynamical system, i.e., the mathematical space formed by the dependent variables of a system. An extended phase space is the Cartesian product of the phase space with the independent variable, which is often time.

Phenomenon A collection of percepts to which relationships are assigned.

Phenotype The observable physical or biochemical characteristics of an organism, as determined by both genetic makeup and environmental influences.

Phosphatase Enzyme that removes phosphate groups from a molecule.

Phosphorylation Addition of a phosphate group to a protein molecule. The resulting conformational change usually activates the protein.

Poisson distribution A discrete probability distribution that expresses the probability of a number of events occurring in a fixed period of time if these events occur with a known average rate and independently of the time since the last event.

Polymer Large molecule synthesized by linking monomers together.

Pore Pores are made of pore-forming proteins, called *porins*. Pores allow small *hydrophilic molecules* (molecules that can transiently bond with water) to pass through the membrane of an organelle.

Positive feedback Feedback that results in amplification or growth of the output signal.

Probability density function A function that describes the relative likelihood for a random variable to occur at a given point in the observation space.

Probability generating function A power series representation (the generating function) of the probability distribution of a random variable.

Probability mass function A function that gives the probability that a discrete random variable is exactly equal to some value.

Probability measure A probability measure is a measure with total measure one; a probability space is a measure space with a probability measure. See also Measure.

Probability space A mathematical construct that models a real-world process (or experiment) consisting of states that occur randomly. A probability space is constructed with a specific kind of situation or experiment in mind.

Product A substance resulting from a chemical reaction. In the context of mathematical operations, a product is a result of multiplication.

Promoter A promoter is a region of DNA that facilitates the transcription of a particular gene. Promoters are typically located near the genes they regulate, on the same strand and upstream.

Propensity Probability per unit time that an elementary reaction in a given state occurs.

Protein A linear polymer of linked amino acids, referred to as a macromolecule and major constituent component of the cell.

Protease, proteinase Enzymes that degrade proteins by splitting internal peptide bonds to produce peptides.

Proteinase inhibitor Small proteins that inhibit various proteinase enzymes. An example is antitrypsin.

Protein kinase Enzyme that transfers the terminal phosphate group of ATP to a specific amino acid of a target protein.

Pseudo steady state A condition in which it is convenient to assume a steady state for portions of a non-steady-state system.

Pumps proteins that actively transport ions and other molecules across cellular and intracellular membranes. Pumps can work against an electrochemical gradient.

Quasi steady state State of a system's component where its rate of change is negligible in comparison to other system components. Then, its rate of change is assumed to be approximately zero.

Randomness Lack of any specific pattern, in terms of predictability, in an observed phenomenon.

Random experiment An experiment, trial, or observation that can be repeated numerous times under the same conditions.

Random number A number that occurs in a sequence such that two conditions are met: (1) the values are distributed over a defined interval or set according to some rule, and (2) it is impossible to predict future values based on past or present ones.

Random process A description of real or simulated data for which the behavior is or appears unpredictable.

Random variable Can be thought of as an unknown value that may change every time it is inspected.

Random vector A finite-dimensional formal vector of random variables.

Rare event Uncommon event.

Rate law An equation that links the reaction rate with concentrations or pressures of reactants and constant parameters (normally rate coefficients and partial reaction orders).

Reactant A chemical substance that is present at the start of a chemical reaction.

Reaction count Number of reactions of a certain type that occur in a specific time frame.

Reaction network Complete set of chemical reactions with their regulatory interactions.

Reaction pathway Subset of chemical reactions in a reaction network leading to a specific product.

Realization The realization of a random variable is the value that is actually observed (what actually happened).

Receptor Participating entity that binds to a specific physical entity and initiates the response to that physical entity.

Recurrence relation An equation that recursively defines a sequence: each term of the sequence is defined as a function of the preceding terms.

Regulation The maintenance of a regular or desirable state, making a system robust against perturbations. See also homeostasis and control.

Replication The process by which genetic material, a single-cell organism, or a virus reproduces or makes a copy of itself.

Repressor Protein that binds to a specific region of DNA to prevent transcription of an adjacent gene.

Reversible Of or relating to a process, such as a chemical reaction or a phase change, in which the system undergoing the process can be returned to its original state.

Right-tail distribution function See complementary cumulative distribution function.

Sample point A possible result of an experiment, represented as a point.

Sample space The set of all possible outcomes.

S-phase Short for synthesis phase, a period in the cell cycle during interphase, between the G1 phase and the G2 phase.

Sample space The set of possible outcomes in a statistical experiment.

Sensitivity analysis A tool used during the model-making process to determine the quantity of variation in the observable quantities that can be attributed to variation in each input parameter.

Signal amplification Any process by which a cell converts one kind of signal or stimulus into another.

Signaling, signal transduction A process by which signals are relayed through biochemical reactions.

SimBiology A Matlab toolbox that provides an integrated environment for modeling biological processes, simulating the dynamic behavior of these processes, and analyzing the model with simulation and experimental data.

Standard deviation A statistic used as a measure of the dispersion or variation in a distribution, equal to the square root of the arithmetic mean of the squares of the deviations from the arithmetic mean.

State space The collection of all possible states of a system.

Stationary increments Increments such that the probability distribution of any increment during a time interval depends only on the interval length and not on the time point of measurement; increments with equally long time intervals are identically distributed.

Steady state A system is said to be in a steady state if the recently observed behavior will continue in the future. A steady state is usually obtained by setting all time derivatives to zero, a justifiable assumption for equilibrated systems.

Stochastic differential equation A differential equation in which one or more of the terms is a stochastic process, thus resulting in a solution that is itself a stochastic process.

Stochastic model A model that incorporates random fluctuations in model parameters or model structure.

Stochastic process The counterpart to a deterministic process (or deterministic system). Instead of dealing with only one possible reality of how the process might evolve over time (as is the case, for example, for solutions of an ordinary differential equation), in a stochastic or random process there is some indeterminacy in its future evolution described by probability distributions.

Stochastic simulation algorithm Generates a statistically correct trajectory (possible solution) of a stochastic equation.

Stoichiometric coefficient The stoichiometric coefficient represents the degree to which a chemical species participates in a reaction. It corresponds to the number of molecules of a reactant that are consumed or produced with each occurrence of a reaction event.

Stoichiometry Calculation of the quantities of reactants and products in a chemical reaction.

Subcellular Smaller than an ordinary cell; below cellular level or scope.

Substrate Molecule that is acted upon by an enzyme. The substrate binds with the enzyme's active site, and the enzyme catalyzes a chemical reaction involving the substrate.

System A collection of objects and a relation among these objects.

System theory Theory of the laws that govern the behavior and interaction of systems.

Taylor expansion A representation of a function as an infinite sum of terms calculated from the values of its derivatives at a single point.

Thermal equilibrium A state in which all parts of a system are at the same temperature.

Trajectory The solution of a set of differential equations, synonymous with the phrase "phase curve".

Transcription Process through which a DNA sequence is copied to produce a complementary RNA.

Transcription factor A protein that binds to specific DNA sequences, thereby controlling the movement (or transcription) of genetic information from DNA to mRNA. Transcription factors perform this function alone or with other proteins in a complex, by promoting (as an activator) or blocking (as a repressor) the recruitment of RNA polymerase (the enzyme that performs the transcription of genetic information from DNA to RNA) to specific genes.

Translation Process in which a polypeptide chain is produced from a messenger RNA.

Transition Passage from one form, state, style, or place to another.

Transition probability Conditional probability concerning a discrete Markov chain giving the probabilities of change from one state to another.

Transition state theory Explains the reaction rates of elementary chemical reactions. The theory assumes a special type of chemical equilibrium (quasi equilibrium) between reactants and activated transition state complexes.

Uncertainty The lack of certainty. A state of having limited knowledge in which it is impossible to describe exactly the existing state or future outcome among more than one possible outcome.

Uniform distribution A family of probability distributions such that for each member of the family, all intervals of the same length on the distribution's support are equally probable.

Variance The expected, or mean, value of the square of the deviation of that variable from its expected value or mean.

Wiener process A continuous-time stochastic process named in honor of Norbert Wiener. It is often called Brownian motion.

Index